How to Think Like a Mathematician

A Companion to Undergraduate Mathematics

KEVIN HOUSTON

University of Leeds

CAMBRIDGE UNIVERSITY PRESS
Cambridge, New York, Melbourne, Madrid, Cape Town, Singapore, São Paulo, Delhi

Cambridge University Press
The Edinburgh Building, Cambridge CB2 8RU, UK

Published in the United States of America by Cambridge University Press, New York

www.cambridge.org
Information on this title: www.cambridge.org/9780521895460

1005719234

First published 2009

Printed in the United Kingdom at the University Press, Cambridge

A catalogue record for this publication is available from the British Library

Library of Congress Cataloging-in-Publication Data

Houston, Kevin, 1968–
How to think like a mathematician : a companion to undergraduate mathematics / Kevin Houston.
        p.   cm.
    Includes index.
1. Mathematics–Study and teaching (Higher)–United States.
I. Title.
    QA13.H68 2009
    510–dc22

                                                                          2008034663

ISBN 978-0-521-89546-0 hardback
ISBN 978-0-521-71978-0 paperback

To Mum and Dad – Thanks for everything.

Contents

Preface

Question: How many months have 28 days?
Mathematician's answer: All of them.

The power of mathematics

Mathematics is the most powerful tool we have. It controls our world. We can use it to put men on the moon. We use it to calculate how much insulin a diabetic should take. It is hard to get right.

And yet. And yet . . . And yet people who use or like mathematics are considered geeks or nerds.[1] And yet mathematics is considered useless by most people – throughout history children at school have whined 'When am I ever going to use this?'

Why would anyone want to become a mathematician? As mentioned earlier mathematics is a very powerful tool. Jobs that use mathematics are often well-paid and people do tend to be impressed. There are a number of responses from non-mathematicians when meeting a mathematician, the most common being 'I hated maths at school. I wasn't any good at it', but another common response is 'You must be really clever.'

The concept

The aim of this book is to divulge the secrets of how a mathematician actually thinks. As I went through my mathematical career, there were many instances when I thought, 'I wish someone had told me that earlier.' This is a collection of such advice. Well, I hope it is more than such a collection. I wish to present an attitude – a way of thinking and doing mathematics that works – not just a collection of techniques (which I will present as well!)

If you are a beginner, then studying high-level mathematics probably involves using study skills new to you. I will not be discussing generic study skills necessary for success – time management, note taking, exam technique and so on; for this information you must look elsewhere.

I want you to be able to think like a mathematician and so my aim is to give you a book jam-packed with practical advice and helpful hints on how to acquire skills specific to

---

[1] Add your own favourite term of abuse for the intelligent but unstylish.

thinking like a mathematician. Some points are subtle, others appear obvious when you have been told them. For example, when trying to show that an equation holds you should take the most complicated side and reduce it until you get to the other side (page 143). Some advice involves high-level mathematical thinking and will be too sophisticated for a beginner – so don't worry if you don't understand it all immediately.

How to use this book

Each part has a different style as it deals with a different idea or set of ideas. The book contains a lot of information and, like most mathematics books, you can't read it like a novel in one sitting.

Some friendly advice

And now for some friendly advice that you have probably heard before – but is worth repeating.

- *It's up to you* – Your actions are likely to be the greatest determiner of the outcome of your studies. Consider the ancient proverb: The teacher can open the door, but you must enter by yourself.
- *Be active* – Read the book. Do the exercises set.
- *Think for yourself* – Always good advice.
- *Question everything* – Be sceptical of all results presented to you. Don't accept them until you are sure you believe them.
- *Observe* – The power of Sherlock Holmes came not from his deductions but his observations.
- *Prepare to be wrong* – You will often be told you are wrong when doing mathematics. Don't despair; mathematics is hard, but the rewards are great. Use it to spur yourself on.
- *Don't memorize – seek to understand* – It is easy to remember what you truly understand.
- *Develop your intuition* – But don't trust it completely.
- *Collaborate* – Work with others, if you can, to understand the mathematics. This isn't a competition. Don't merely copy from them though!
- *Reflect* – Look back and see what you have learned. Ask yourself how you could have done better.

To instructors and lecturers – a moment of your valuable time

One of my colleagues recently complained to me that when a student is given a statement of the form A implies B to prove their method of proof is generally wholly inadequate. He jokingly said, the student assumes A, works with that for a bit, uses the fact that B is true and so concludes that A is true. How can it be that so many students have such a hard time constructing logical arguments that form the backbone of proofs?

I wish I had an answer to this. This book is an attempt at an answer. It is not a theoretical manifesto. The ideas have been tried and tested from years of teaching to improve mathematical thinking in my students. I hope I have provided some good techniques to get them onto the path of understanding.

If you want to use this book, then I suggest you take your favourite bits or pick some techniques that you know your own students find hard, as even I think that students cannot swallow every piece of advice in this book in a single course. One aim in my own teaching is to be inspirational to students. Mathematics should be exciting. If the students feel this excitement, they are motivated to study and, as in the proverb quoted above, will enter by themselves. I aim to make them free to explore, give them the tools to climb the mountains, and give them their own compasses so they can explore other mathematical lands. Achieving this is hard, as you know, and it is often not lack of time, resources, help from the university or colleagues that is the problem. Often, through no fault of their own, it is the students themselves. Unfortunately, they are not taught to have a questioning nature, they are taught to have an answering nature. They expect us to ask questions and for them to give the answers because that is they way they have been educated. This book aims to give them the questions they need to ask so they don't need me anymore.

I'd just like to thank . . .

This book has had a rather lengthy genesis and so there are many people to thank for influencing me or my choice of contents. Some of the material appeared in a booklet of the same name, given to all first-year Mathematics students at the University of Leeds, and so many students and staff have given their opinions on it over the years. The booklet was available on the web, and people from around the world have sent unsolicited comments. My thanks go to Ahmed Ali, John Bibby, Garth Dales, Tobias Gläßer, Chris Robson, Sergey Klokov, Katy Mills, Mike Robinson and Rachael Smith, and to students at the University of Leeds and at the University of Warwick who were first subjected to my wild theories and experiments (and whose names I have forgotten). Many thanks to David Franco, Margit Messmer, Alan Slomson and Maria Veretennikova for reading a preliminary draft. Particular thanks to Margit and Alan with whom I have had many fruitful discussions. My thanks to an anonymous referee and all the people at the Cambridge University Press who were involved in publishing this book, in particular, Peter Thompson.

Lastly, I would like to thank my gorgeous wife Carol for putting up with me while I was writing this book and for putting the sunshine in my life.

PART I

Study skills for mathematicians

CHAPTER 1

Sets and functions

Everything starts somewhere, although many physicists disagree.
Terry Pratchett, *Hogfather*, 1996

To think like a mathematician requires some mathematics to think about. I wish to keep the number of prerequisites for this book low so that any gaps in your knowledge are not a drag on understanding. Just so that we have some mathematics to play with, this chapter introduces sets and functions. These are very basic mathematical objects but have sufficient abstraction for our purposes.

A set is a collection of objects, and a function is an association of members of one set to members of another. Most high-level mathematics is about sets and functions between them. For example, calculus is the study of functions from the set of real numbers to the set of real numbers that have the property that we can differentiate them. In effect, we can view sets and functions as the mathematician's building blocks.

While you read and study this chapter, think about *how* you are studying. Do you read every word? Which exercises do you do? Do you, in fact, do the exercises? We shall discuss this further in the next chapter on reading mathematics.

Sets

The set is the fundamental object in mathematics. Mathematicians take a set and do wonderful things with it.

Definition 1.1

*A **set** is a well-defined collection of objects.*[1]
   *The objects in the set are called the **elements** or **members** of the set.*

We usually define a particular set by making a list of its elements between brackets. (We don't care about the ordering of the list.)

---

[1] The proper mathematical definition of set is much more complicated; see almost any text book on set theory. This definition is intuitive and will not lead us into many problems. Of course, a pedant would ask what does well-defined mean?

If x is a member of the set X, then we write $x \in X$. We read this as 'x is an element (or member) of X' or 'x is in X'.[2] If x is not a member, then we write $x \notin X$.

Examples 1.2

(i) The set containing the numbers 1, 2, 3, 4 and 5 is written $\{1, 2, 3, 4, 5\}$. The number 3 is an element of the set, i.e. $3 \in \{1, 2, 3, 4, 5\}$, but $6 \notin \{1, 2, 3, 4, 5\}$. Note that we could have written the set as $\{3, 2, 5, 4, 1\}$ as the order of the elements is unimportant.

(ii) The set $\{$dog, cat, mouse$\}$ is a set with three elements: dog, cat and mouse.

(iii) The set $\{1, 5, 12, \{$dog, cat$\}, \{5, 72\}\}$ is the set containing the numbers 1, 5, 12 and the sets $\{$dog, cat$\}$ and $\{5, 72\}$. Note that sets can contain sets as members. Realizing this now can avoid a lot of confusion later.

It is vitally important to note that $\{5\}$ and 5 are not the same. That is, we must distinguish between being a set and being an element of a set. Confusion is possible since in the last example we have $\{5, 72\}$, which is a set in its own right but can also be thought of as an element of a set, i.e. $\{5, 72\} \in \{1, 5, 12, \{dog, cat\}, \{5, 72\}\}$.

Let's have another example of a set created using sets.

Example 1.3

The set $X = \{1, 2, dog, \{3, 4\}, $mouse$\}$ has five elements. It has the the four elements, 1, 2, dog, mouse; and the other element is the set $\{3, 4\}$. We can write $1 \in X$, and $\{3, 4\} \in X$. It is vitally important to note that $3 \notin X$ and $4 \notin X$, i.e. the numbers 3 and 4 are not members of X, the set $\{3, 4\}$ is.

Some interesting sets of numbers

Let's look at different types of numbers that we can have in our sets.

Natural numbers

The set of **natural numbers** is $\{1, 2, 3, 4, \dots\}$ and is denoted by $\mathbb{N}$. The dots mean that we go on forever and can be read as 'and so on'.

Some mathematicians, particularly logicians, like to include 0 as a natural number. Others say that the natural numbers are the counting numbers and you don't start counting with zero (unless you are a computer programmer). Furthermore, how natural is a number that was not invented until recently?

On the other hand, some theorems have a better statement if we take $0 \in \mathbb{N}$. One can get round the argument by specifying that we are dealing with non-negative integers or positive integers, which we now define.

---

[2] Of course, to distinguish the x and X we read it out loud as 'little x is an element of capital X.'

Integers

The set of **integers** is $\{\ldots, -4, -3, -2, 0, 1, 2, 3, 4, \ldots\}$ and is denoted by $\mathbb{Z}$. The $\mathbb{Z}$ symbol comes from the German word Zahlen, which means number. From this set it is easy to define the **non-negative integers**, $\{0, 1, 2, 3, 4, \ldots\}$, often denoted $\mathbb{Z}^{+}$. Note that all natural numbers are integers.

Rational numbers

The set of **rational numbers** is denoted by $\mathbb{Q}$ and consists of all fractional numbers, i.e. $x \in \mathbb{Q}$ if x can be written in the form p/q where p and q are integers with $q \neq 0$. For example, $1/2, 6/1$ and $80/5$. Note that the representation is not unique since, for example, $80/5 = 16/1$. Note also that all integers are rational numbers since we can write $x \in \mathbb{Z}$ as $x/1$.

Real numbers

The **real numbers**, denoted $\mathbb{R}$, are hard to define rigorously. For the moment let us take them to be any number that can be given a decimal representation (including infinitely long representations) or as being represented as a point on an infinitely long number line.

The real numbers include all rational numbers (hence integers, hence natural numbers). Also real are π and e, neither of which is a rational number.[3] The number $\sqrt{2}$ is not rational as we shall see in Chapter 23.

The set of real numbers that are not rational are called **irrational numbers**.

Complex numbers

We can go further and introduce **complex numbers**, denoted $\mathbb{C}$, by pretending that the square root of -1 exists. This is one of the most powerful additions to the mathematician's toolbox as complex numbers can be used in pure and applied mathematics. However, we shall not use them in this book.

More on sets

The empty set

The most fundamental set in mathematics is perhaps the oddest – it is the set with no elements!

---

[3] The proof of these assertions are beyond the scope of this book. For π see Ian Stewart, *Galois Theory*, 2nd edition, Chapman and Hall 1989, p. 62 and for e see Walter Rudin, *Principles of Mathematical Analysis*, 3rd edition, McGraw-Hill 1976, p. 65.

Definition 1.4

*The set with no elements is called the **empty set** and is denoted ∅.*

It may appear to be a strange object to define. The set has no elements so what use can it be? Rather surprisingly this set allows us to build up ideas about counting. We don't have time to explain fully here but this set is vital for the foundations of mathematics. If you are interested, see a high level book on set theory or logic.

Example 1.5

The set {∅} is the set that contains the empty set. This set has one element. Note that we can then write ∅ ∈ {∅}, but we *cannot* write ∅ ∈ ∅ as the empty set has, by definition, no elements.

Definition 1.6

*Two sets are **equal** if they have the same elements. If set X equals set Y, then we write $X = Y$. If not we write $X \neq Y$.*

Examples 1.7

(i)   The sets {5, 7, 15} and {7, 15, 5} are equal, i.e. {5, 7, 15} = {7, 15, 5}.
(ii) The sets {1, 2, 3} and {2, 3} are not equal, i.e. {1, 2, 3} ≠ {2, 3}.
(iii) The sets {2, 3} and {{2}, 3} are not equal.
(iv) The sets ℝ and ℕ are not equal.

Note that, as used in the above, if we have a symbol such as = or ∈, then we can take the opposite by drawing a line through it, such as ≠ and ∉.

Definition 1.8

*If the set X has a finite number of elements, then we say that X is a **finite set**. If X is finite, then the number of elements is called the **cardinality** of X and is denoted $|X|$.*

If X has an infinite number of elements, then it becomes difficult to define the cardinality of X. We shall see why in Chapter 30. Essentially it is because there are different sizes of infinity! For the moment we shall just say that the cardinality is undefined for infinite sets.

Examples 1.9

(i) The set {∅, 3, 4, cat} has cardinality 4.
(ii) The set {∅, 3, {4, cat} } has cardinality 3.

Exercises 1.10

What is the cardinality of the following sets?

(i)   {1, 2, 5, 4, 6}   (ii)   {π, 6, {π, 5, 8, 10}}
(iii) {π, 6, {π, 5, 8, 10}, {dog, cat, {5}}}   (iv)   ∅
(v)   ℕ   (vi)   {dog, ∅}
(vii) {∅, {∅, {∅}}}   (viii) {∅, {20, π, {∅}}, 14}

Now we come to another crucial definition, that of being a subset.

Definition 1.11

*Suppose X is a set. A set Y is a **subset** of X if every element of Y is an element of X. We write $Y \subseteq X$.*

This is the same as saying that, if $x \in Y$, then $x \in X$.

Examples 1.12

(i)   The set $Y = \{1, \{3, 4\}, \text{mouse}\}$ is a subset of $X = \{1, 2, \text{dog}, \{3, 4\}, \text{mouse}\}$.
(ii) The set of even numbers is a subset of $\mathbb{N}$.
(iii) The set $\{1, 2, 3\}$ is not a subset of $\{2, 3, 4\}$ or $\{2, 3\}$.
(iv) For any set X, we have $X \subseteq X$.
(v)   For any set X, we have $\emptyset \subseteq X$.

Remark 1.13

It is vitally important that you distinguish between being an *element* of a set and being a *subset* of a set. These are often confused by students. If $x \in X$, then $\{x\} \subseteq X$. Note the brackets. Usually, and I stress usually, if $x \in X$, then $\{x\} \notin X$, but sometimes $\{x\} \in X$, as the following special example shows.

Example 1.14

Consider the set $X = \{x, \{x\}\}$. Then $x \in X$ and $\{x\} \subseteq X$ (the latter since $x \in X$) but we also have $\{x\} \in X$.

Therefore we *cannot* state any simple rule such as 'if $a \in A$, then it would be wrong to write $a \subseteq A$', and vice versa.

If you felt a bit confused by that last example, then go back and think about it some more, until you really understand it. This type of precision and the nasty examples that go against intuition, and prevent us from using simple rules, are an important aspect of high-level mathematics.

Definition 1.15

*A subset Y of X is called a **proper subset** of X if Y is not equal to X. We denote this by $Y \subset X$. Some people use $Y \subsetneq X$ for this.*

Examples 1.16

(i) $\{1, 2, 5\}$ is a proper subset of $\{-6, 0, 1, 2, 3, 5\}$.
(ii) For any set X, the subset X is not a proper subset of X.
(iii) For any set $X \neq \emptyset$, the empty set $\emptyset$ is a proper subset of X. Note that, if $X = \emptyset$, then the empty set $\emptyset$ is *not* a proper subset of X.
(iv) For numbers, we have $\mathbb{N} \subset \mathbb{Z} \subset \mathbb{Q} \subset \mathbb{R} \subset \mathbb{C}$.

Note that we can use the symbols $\not\subseteq$ to denote 'not a subset of' and $\not\subset$ to denote 'not a proper subset of'.

Now let's consider where the notation came from. It is is obvious that for a finite set the two statements

$$\text{If } X \subseteq Y, \text{ then } |X| \leq |Y|,$$

and

$$\text{If } X \subset Y, \text{ then } |X| < |Y|$$

are true. So $\subseteq$ is similar to $\leq$ and $\subset$ is similar to $<$ as concepts and not just as symbols.

An important remark to make here is that not all mathematicians distinguish between $\subseteq$ and $\subset$; some use only $\subset$ and use it to mean 'subset of'. However, I feel the use of $\subseteq$ is far better as it allows us to distinguish between a subset and a proper subset. Imagine what the two statements above would look like if we didn't. They wouldn't be so clear and one wouldn't be true! Or, to see what I mean, imagine what would happen if mathematicians always used $<$ instead of $\leq$.

Defining sets

We can define sets using a different notation: $\{x \mid x \text{ satisfies property } P\}$. The symbol '$\mid$' is read as 'such that'. Sometimes the colon ':' is used in place of '$\mid$'.

Examples 1.17

(i)   The set $\{x \mid x \in \mathbb{N} \text{ and } x < 5\}$ is equal to $\{1, 2, 3, 4\}$. We read the set as 'x such that x is in $\mathbb{N}$ and x is less than 5'.

(ii) The set $\{x \mid 5 \leq x \leq 10\}$ is the set of numbers between 5 and 10. Here we follow the convention that we assume that x is a real number. This is a bad convention as it allows writers to be sloppy, so we should try to avoid using it. Hence, we can also specify some restriction on the x before the $\mid$ sign, as in the next example.

(iii) The set $\{x \in \mathbb{N} \mid 5 \leq x \leq 10\}$ is the set of natural numbers from 5 to 10 inclusive. That is, the set $\{5, 6, 7, 8, 9, 10\}$.

(iv) It is common to use the notation $[a, b]$ for the set $\{x \in \mathbb{R} \mid a \leq x \leq b\}$ and (a, b) for the set $\{x \in \mathbb{R} \mid a < x < b\}$.

Note that (a, b) can also mean the pair of numbers a and b.

We can also describe sets in the following way $\{x^2 \mid x \in \mathbb{N}\}$ is the set of numbers $\{1, 4, 9, 16, \dots\}$. There are many possibilities for describing sets so we will not detail them all as it will usually be obvious what is intended.

Operations on sets

In mathematics we often make a definition of some object, for example a set, and then we find ways of creating new ones from old ones, for example we take subsets of sets. We now come to two ways of creating new from old: the union and intersection of sets.

Definition 1.18

*Suppose that X and Y are two sets. The **union** of X and Y, denoted $X \cup Y$, is the set consisting of elements that are in X or in Y or in both. We can define the set as $X \cup Y = \{x \mid x \in X \text{ or } x \in Y\}$.*

Examples 1.19

(i) The union of $\{1, 2, 3, 4\}$ and $\{2, 4, 6, 8\}$ is $\{1, 2, 3, 4, 6, 8\}$.

(ii) The union of $\{x \in \mathbb{R} \mid x < 5\}$ and $\{x \in \mathbb{Z} \mid x < 8\}$ is $\{x \in \mathbb{R} \mid x \leq 5$, or $x = 6$ or $x = 7\}$.

Exercises 1.20

(i) Let $X = \{1, 2, 3, 4, 5\}$ and $Y = \{-1, 1, 3, 5, 7\}$. Find $X \cup Y$.

(ii) What is $\mathbb{Z} \cup \mathbb{Z}$?

Definition 1.21

*Suppose that X and Y are two sets. The **intersection** of X and Y, denoted $X \cap Y$, is the set consisting of elements that are in X and in Y. We can define the set as $X \cap Y = \{x \mid x \in X$ and $x \in Y\}$.*

Examples 1.22

(i) The intersection of $\{1, 2, 3, 4\}$ and $\{2, 4, 6, 8\}$ is $\{2, 4\}$.

(ii) The intersection of $\{-1, -2, -3, -4, -5\}$ and $\mathbb{N}$ is $\emptyset$.

Exercises 1.23

(i) Find $X \cap Y$ for the following:

    (a) $X = \{x \in \mathbb{R} \mid 0 \leq x < 6\}$ and $Y = \{x \in \mathbb{Z} \mid -\pi \leq x \leq 7\}$,

    (b) $X = \{0, 2, 4, 6, 8\}$ and $Y = \{1, 3, 5, 7, 9\}$,

    (c) $X = \mathbb{Q}$ and $Y = \{0, 1, \pi, 5\}$.

(ii) Find $\mathbb{Z} \cap \mathbb{Z}$, $\mathbb{Z} \cap \emptyset$, and $\mathbb{Z} \cap \mathbb{R}$.

We will use these definitions in later chapters to give examples of proofs, for example to show statements such as $X \cap (Y \cup Z) = (X \cap Y) \cup (X \cap Z)$ are true.

Exercise 1.24

Find the union and intersection of $\{x \in \mathbb{R} \mid x > 7\}$ and $\{x \in \mathbb{N} \mid x > 5\}$.

Definition 1.25

*The **difference** of X and Y, denoted $X \backslash Y$, is the set of elements that are in X but not in Y. That is, we take elements of X and discard those that are also in Y. We do not require that Y is a subset of X. If Y is defined as a subset of X, then we often call $X \backslash Y$ the **complement** of Y in X and denote this by Y^c.*

Examples 1.26

(i) Let $X = \{1, 2, 3, \text{dog}, \text{cat}\}$ and let $Y = \{3, \text{cat}, \text{mouse}\}$. Then $X \backslash Y = \{1, 2, \text{dog}\}$.

(ii) Let $X = \mathbb{R}$ and $Y = \mathbb{Z}$, then

$$X \backslash Y = \cdots \cup (-3, -2) \cup (-2, -1) \cup (-1, 0) \cup (0, 1) \cup (1, 2) \cup \cdots$$

Products of sets

Here's another example of mathematicians creating new objects from old ones.

Definition 1.27

*Let X and Y be two sets. The **product** of X and Y, denoted $X \times Y$ is the set of all possible pairs (x, y) where $x \in X$ and $y \in Y$, i.e.*

$$X \times Y = \{(x, y) \mid x \in Y \text{ and } y \in Y\}.$$

Note that here (x, y) denotes a pair and has nothing to do with Example 1.17(iv).

Examples 1.28

(i) Let $X = \{0, 1\}$ and $Y = \{1, 2, 3\}$. Then $X \times Y$ has six elements:

$$X \times Y = \{(0, 1), (0, 2), (0, 3), (1, 1), (1, 2), (1, 3)\}.$$

(ii) The set $\mathbb{R} \times \mathbb{R}$ is denoted $\mathbb{R}^2$. The set $(\mathbb{R} \times \mathbb{R}) \times \mathbb{R}$ is denoted $\mathbb{R}^3$. This is because its elements can be given by triples of real numbers, i.e. its elements are of the form (x, y, z) where x, y and z are real numbers.

Note that $X \times Y$ is not a subset of either X or Y.

Maps and functions

We have defined sets. Now we make a definition for relating elements of sets to elements of other sets.

Definition 1.29

*Suppose that X and Y are sets. A **function** or **map** from X to Y is an association between the members of the sets. More precisely, for every element of X there is a unique element of Y.*

*If f is a function from X to Y, then we write $f : X \to Y$, and the unique element in Y associated to x is denoted $f(x)$. This element is called the **value of x under** f or called a **value** of f. The set X is called the **source** (or **domain**) of f and Y is called the **target** (or **codomain**) of f.*

To describe a function f we usually use a formula to define $f(x)$ for every x and talk about applying f to elements of a set, or to a set.

A schematic picture is shown in Figure 1.1. Note that every element of X has to be associated to one in Y but not vice versa and that two distinct elements of X may map to the same one in Y.

Examples 1.30

(i) Let $f : \mathbb{Z} \to \mathbb{Z}$ be defined by $f(x) = x^2$ for all $x \in \mathbb{Z}$. Then the value of x under f is the square of x. Note that there are elements in the target which are not values of f. For example -1 is not a value since there is no integer x such that $x^2 = -1$.

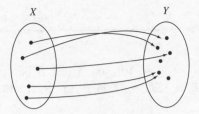

Figure 1.1 A function from X to Y

(ii) Let $f : \mathbb{R} \to \mathbb{R}$ be given by $f(x) = 0$. Then the only value of f is 0.

(iii) The cardinality of a set is a function on the set of finite sets. That is $|\ | :$ Finite Sets $\to$ $\{0\} \cup \mathbb{N}$. Note that we need 0 in the codomain as the set could be the empty set.

(iv) The **identity map** on X is the map id $: X \to X$ given by $\mathrm{id}(x) = x$ for all $x \in X$.

Having a formula does not necessarily define a function, as the next example shows.

Example 1.31

The formula $f(x) = 1/(x-1)$ does not define a function from $\mathbb{R}$ to $\mathbb{R}$ as it is not defined for $x = 1$.

We can rescue this example by restricting the source to $\mathbb{R}$ without the element 1. That is, define $X = \{x \in \mathbb{R} \mid x \neq 1\}$, then $f : X \to \mathbb{R}$ defined by $f(x) = 1/(x-1)$ is a function.

Polynomials provide a good source of examples of functions.

Examples 1.32

(i) Let $f : \mathbb{R} \to \mathbb{R}$ be given by $f(x) = x^2 + 2x + 3$. Notice again that, although the target is all of $\mathbb{R}$, not every element of the target is a value of f. For example there is no x such that $f(x) = -2$. This is something you can check by attempting to solve $x^2 + 2x + 3 = -2$.

(ii) More generally, from a polynomial we can define a function $f : \mathbb{R} \to \mathbb{R}$ by defining

$$f(x) = a_n x^n + a_{n-1} x^{n-1} + \cdots + a_1 x + a_0$$

for some real numbers $a_0, \ldots, a_n$ and a real variable x.

(iii) Suppose that $f : \mathbb{R} \to \mathbb{R}$ can be differentiated, for example a polynomial. Then the derivative, denoted f', is a function.

Exercises 1.33

(i) Find the largest domain that makes $f(x) = x/(x^2 - 5x + 3)$ a function.

(ii) Find the largest domain that makes $f(x) = (x^3 + 2)/(x^2 + x + 2)$ a function.

(iii) Construct an example of a polynomial so that its graph goes through the points $(-1, 5)$ and $(3, -2)$.

Exercises

Exercises 1.34

(i) Let $X = \{x \in \mathbb{Z} \mid 0 \leq x \leq 10\}$ and A and B be subsets such that $A = \{0, 2, 4, 6, 8, 10\}$ and $B = \{2, 3, 5, 7\}$. Find $A \cap B$, $A \cup B$, $A \backslash B$, $B \backslash A$, $A \times B$, $X \times A$, A^c, and B^c.

(ii) Find the union and intersection of $\{x \in \mathbb{R} \mid x^2 - 9x + 14 = 0\}$ and $\{y \in \mathbb{Z} \mid 3 \leq y < 10\}$.

(iii) Suppose that A, B and C are subsets of X. Use examples of these sets to investigate the following:

(a) $(A \cap B) \cup (A \cap C)$ and $A \cap (B \cup C)$,

(b) $(A \cup B) \cap (A \cup C)$ and $A \cup (B \cap C)$,

(c) $(A \cup B)^c$ and $A^c \cap B^c$,

(d) $(A \cup B)^c$ and $A^c \cup B^c$,

(e) $(A \cap B)^c$ and $A^c \cup B^c$,

(f) $(A \cap B)^c$ and $A^c \cap B^c$.

Do you notice anything?

(iv) A **Venn diagram** is useful way of representing sets. If A is a subset of X, then we can draw the following in the plane:

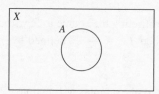

In fact, the precise shape of A is unimportant but we often use a circle. If B is another subset, then we can draw B in the diagram as well. In the following we have shaded the intersection $A \cap B$.

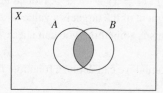

(a) Draw a Venn diagram for the case that A and B have no intersection.

(b) Draw Venn diagrams and shade the sets $A \cup B$, A^c, and $(A \cap B)^c$.

(c) Draw three (intersecting) circles to represent the sets A, B and C. Shade in the intersection $A \cap B \cap C$.

(d) Using exercise (iii) construct Venn diagrams and shade in the relevant sets.

(v) Analyse how you approached the reading of this chapter.

(a) If you had not met the material in this chapter before, then did you attempt to understand everything?

(b) If you had met the material before, did you check to see that I had not made any mistakes?

Summary

- ▶ A set is a well-defined collection of objects.
- ▶ The empty set has no elements.
- ▶ The cardinality of a finite set is the number of elements in the set.
- ▶ The set Y is a subset of X if every element of Y is in X.
- ▶ A subset Y of X is a proper subset if it is not equal to X.
- ▶ The union of X and Y is the collection of elements that are in X or in Y.
- ▶ The intersection of X and Y is the collection of elements that are in X and in Y.
- ▶ The product of X and Y is the set of all pairs (x, y) where $x \in X$ and $y \in Y$.
- ▶ A function assigns elements of one set to another.

CHAPTER 2

Reading mathematics

Don't believe everything you read.
Anon

Obviously you can read and probably you have been taught reading skills for academic purposes as part of a study skills course. Unfortunately, mathematics has some special subtleties which often get missed in classes or books on how to study. For example, speed reading is recommended as a valuable tool for learning in many subjects. In mathematics, however, this is not a good method. Mathematics is rarely overwritten; there are few superfluous adjectives, every word and symbol is important and their omission would render the material incomprehensible or incorrect.

The hints and tips here, which include a systematic method for breaking down reading into digestible pieces, are practical suggestions, not a rigid list of instructions. The main points are the following:

- You should be flexible in your reading habits – read many different treatments of a subject.
- Reading should be a dynamic process – you should be an active, not passive, reader, working with a pen and paper at hand, checking the text and verifying what the author asserts is true.

The last point is where thinking mathematically diverges from thinking in many other subjects, such as history and sociology. You really do need to be following the details as you go along – check them. In history (assuming you don't have a time machine) you can't check that Caesar invaded Britain in 55 BC, you can only check what other people have claimed he did. In mathematics you really can, and should, verify the truth.

The following applies to reading lecture notes and web pages, not just to books, but to make a simpler exposition I shall refer only to books. Tips on specific situations, such as reading a definition, theorem[1] or proof are given in later chapters.

---

[1] A theorem is a mathematical statement that is true. Theorems will be discussed in greater detail in Part III.

Basic reading suggestions

Read with a purpose

The primary goal of reading is to learn, but we may be aiming to consolidate, clarify, or find an overview of some material.

Before reading decide what you want from the text. The goal may be as specific as learning a particular definition or how to solve a certain type of problem, such as integrating products. Whatever the reason, it is important that you do not start reading in the vague hope that everything will become clear.

How did you read the previous chapter of this book? What was your goal? Did you skim through it first to see if you already knew it? Did you want to read it in detail until you were confident that you understood everything? Answering these questions often gives an insight into what you really need to do when reading.

Choose a book at the right level

Some books are not well written and some may be unsuitable for your style of learning. In choosing a book bear in mind two connected points. Every book is written for an audience and a purpose. You may not be the audience, and the book's purpose, which might be to teach a novice or to be used as a reference for experts, may not match the purpose you require.

On the other hand do not reject advanced books totally since early chapters in a book often contain a useful summary of a subject.

Read with pen and paper at hand

Be active – read with pen and paper at hand.

The first reason for using pen and paper is that you should make notes from what you are reading – in particular, what it means, not what it says – and to record ideas as they occur to you. Don't take notes the first time you read through as you will probably copy too much without a lot of understanding.

The second reason is more important. You can explore theorems and formulas[2] by applying them to examples, draw diagrams such as graphs, solve – and even create your own – exercises. This is an important aspect of thinking like a mathematician. Physicists and chemists have laboratory experiments, mathematicians have these explorations as their experiments.

Reading with pen and paper at this stage excludes the use of fluorescent markers! The general tendency when using such pens is to mark everything, so wait until you need to summarize the text.

---

[2] Rather than use 'formulae', the correct latin plural of formula, I'll use a more natural English plural.

Don't read it like a novel

Do not read mathematics like a novel. You do not have to read from cover to cover or in the sequence presented. It is perfectly acceptable to dip in and out, take what is relevant to your situation, and to jump from page to page. This is perhaps surprising advice as mathematics is often thought of as a linear subject where ideas are built on top of one another. But, trust me, it isn't created in a linear way and it isn't learned in a linear way.

Add to this the fact that the tracks made by the pioneers of the subject have been covered and the presentation has been improved for public consumption, and you can probably see that you will need to skip backwards and forwards through a text.

Besides, it is unlikely that you will understand every detail in one sitting. You might have to read a passage a number of times before its true meaning becomes clear.

A systematic method

We now outline a five-point method for systematically tackling long pieces:

(i)   Skim through and identify what is important.
(ii) Ask questions.
(iii) Read through carefully. You can do statements first and proofs later if you like.
(iv) Be active. This should include checking the text and doing the exercises.
(v)   Reflect.

This is a simplistic system of reading which, though numbered, doesn't need to be slavishly followed in order. You may have to be flexible and jump from section to section depending on the situation.

Skim

First, look briefly through the text to get an overview. Study skills books often advise students to read the start and end of chapters to get the main conclusions. This does not always work in mathematics books as arguments are not usually summarized in this way, but it is worth trying.

Did you do this with Chapter 1? If I were to do this I would say that the main points are sets, numbers, operations on sets such union and intersection, functions and polynomials.

Identify what is important

In a more careful but not too detailed reading, identify the important points. Look for assumptions, definitions, theorems and examples that get used again and again, as these will be the key to illuminating the theory. If the same definition appears repeatedly in statements, it is important – so learn it!

From Chapter 1, for example, the concept of the empty set looks important, as does the necessity of discriminating between $\in$ and $\subseteq$, in particular their subtle difference.

Look for theorems or formulas that allow you to calculate because calculation is an effective way to get into a subject. Stop and reread that last sentence – I think it's one of the most useful pieces of advice given to me. Often when I am stuck trying to understand some theory attempting to calculate makes it clear. Noticing what allows you to calculate is thus very important.

In Chapter 1 the most obvious notion involving calculation was the cardinality of a set. However, there were no theorems involving it. Nonetheless, you should mark it as something that will be of use later because it involves the possibility of calculation. And in fact we look at calculation of cardinality in Chapter 5.

A more general example is the product rule and chain rule, etc. in calculus. These allow us to calculate the derivative of a function without using the definition of derivative (which is hard to work with).

Ask questions

At this stage it is helpful to pose some questions about the text, such as, Why does the theory hinge on this particular definition or theorem? What is the important result that the text is leading up to and how does it get us there? From your questions you can make a detailed list of what you want from the text.

In the last chapter the main point of the text was to lay the groundwork for material we will use later as examples.

Careful reading

It is now that the careful reading is undertaken. This should be systematic and combined with thinking, doing exercises and solving problems.

Reading is more than just reading the words, you must think about what they mean. In particular, ensure that you know the meaning of every word and symbol; if you don't know or have forgotten, then look back and find out.

For example, one needs to read carefully to ensure that the difference between being a set and being an element of a set is truly grasped.

Stop periodically to review

Do not try to read too much in one go. Stop periodically to review and think about the text. Keep thinking about the big picture, where are we going and how is a particular result getting us there?

Read statements first – proofs later

Many mathematical texts are written so that proofs can be ignored on an initial reading. This is not to say that proofs are unimportant; they are at the heart of mathematics, but usually – not always – can be read later. You *must* tackle the proof at some point.

There were no proofs in the previous chapter. Don't worry, we will produce many proofs later in the book.

Check the text

The necessity to check the text is why you need pen and paper at hand. There are two reasons. First, to fill in the gaps left by the writer. Often we meet phrases like 'By a straight-forward calculation' or 'Details are left to the reader'. In that case, do that calculation or produce those details. This really allows you to get inside the theory.

For example, on page 7 in Chapter 1, I stated 'It is obvious that ... if $X \subseteq Y$, then $|X| \leq |Y|$.' Did you check to see that it really was 'obvious'? Did you try some examples? Similarly did you focus on the non-intuitive facts such as the fact that it is possible to have $\{x\} \in X$ and $\{x\} \subseteq X$ at the same time?

The second reason is to see how theorems, formulas, etc. apply. If the text says use Theorem 3.5 or equation Y, then check that Theorem 3.5 can be applied or check what happens to equation Y in this situation. Verify the formulas and so on. Be a sceptic – don't just take the author's word for it.

Do the exercises and problems

Most modern mathematics books have exercises and problems. It is hard to overplay the importance of doing these. Mathematics is an activity. Think of yourself as not studying mathematics, but *doing* mathematics.

Imagine yourself as having a mathematics muscle. It needs exercise to become developed. Passive reading is like watching someone else training with weights; it won't build your muscles – *you* have to do the exercises.

Furthermore, just because you have read something it does not mean you truly understand it. Answering the exercises and problems identifies your misconceptions and misunderstandings. Regularly I hear from students that they can understand a topic; it's just that they can't do the exercises, or can't apply the material. Basically, my rule is: if I can't do the exercises, then I don't understand the topic.

Reflect

In order to understand something fully we need to relate it to what we already know. Is it analogous to something else? For example, note how the $\subseteq$ notation made sense when it was compared with $\leq$ via cardinality. Can you think what intersection and union might be analogous to?

Another question to ask is 'What does this tell us or allow us to do that other work does not?' For example, the empty set allows us to count (something that was not explained but was alluded to in Chapter 1). Functions allow us to connect sets to sets. Cardinality allows us to talk about the relative sizes of sets. So when you meet a topic ask 'What does it allow me to do?'

What to do afterwards

Don't reread and reread – move on

It is unlikely that understanding will come from excessive rereading of a difficult passage. If you are rereading, then it is probably a sign that you are not active – so do some exercises, ask some questions and so on.

If that fails, it is time to look for an alternative approach, such as consulting another book. Ultimately, it is acceptable to give up and move on to the next part; you can always come back.

By moving on, you may encounter difficulty in understanding the subsequent material, but it might clarify the difficult part by revealing something important.

Also mathematics is a subject that requires time to be absorbed by the brain; ideas need to percolate and have time to grow and develop.

Reread

The assertion to reread may seem strange as the previous piece of advice was not to reread. The difference here is that one should come back much later and reread, for example, when you feel that you have learned the material. This often reveals many subtle points missed or gives a clearer overview of the subject.

Write a summary

The material may appear obvious once you have finished reading, but will that be true at a later date? It is a good time to make a summary – written in your own words.

Exercises

Exercises 2.1

(i) Look back at Chapter 1 and analyse how you attempted to read and understand it.

(ii) Find a journal or a science magazine that includes some mathematics in articles, for example *Scientific American*, *Nature*, or *New Scientist*.

   Read an article. What is the aim of the article and who is the audience? How is the maths used? In one sentence what is the aim? Give three main points.

(iii) Find three textbooks of a similar level and within your mathematical ability. Briefly look through the books and decide which is most friendly and explain your reasons why.

(iv) Find three books tackling the same subject. Find a mathematical object in all three book, say, a set. Are the definitions different in the different books? Which is the best definition? Or rather, which is your favourite definition?

   Are any diagrams used to illustrate the concept? What understanding do the diagrams give? How are the diagrams misleading?

Summary

- ▶ Read with a purpose.
- ▶ Read actively. Have pen and paper with you.
- ▶ You do not have to read in sequence but read systematically.
- ▶ Ask questions.
- ▶ Read the definitions, theorems and examples first. The proofs can come later.
- ▶ Check the text by applying formulas etc.
- ▶ Do exercises and problems.
- ▶ Move on if you are stuck.
- ▶ Write a summary.
- ▶ Reflect – What have you learned?

CHAPTER 3

Writing mathematics I

We have a habit in writing articles published in scientific journals to make the work as
finished as possible, to cover up all the tracks, to not worry about the blind alleys or
describe how you had the wrong idea first, and so on.
Richard Feynman, Nobel Lecture, 1966

As a lecturer my toughest initial task in turning enthusiastic students into able mathematicians is to force them (yes, force them) to write mathematics correctly. Their first
submitted assessments tend to be incomprehensible collections of symbols, with no sentences or punctuation. 'What's the point of writing sentences?', they ask, 'I've got the
correct answer. There it is – see, underlined – at the bottom of the page.' I can sympathize
but in mathematics we have to get to the right answer in a rigorous way and we have to
be able to show to others that our method is rigorous.

A common response when I indicate a nonsensical statement in a student's work is 'But
you are a lecturer, you know what I meant.' I have sympathy with this view too, but there
are two problems with it.

(i) If the reader has to use their intelligence to work out what was intended, then
the student is getting marks because of the reader's intelligence, not their own
intelligence.[1]

(ii) This second point is perhaps more important for students. Sorting through a jumble
of symbols and half-baked poorly expressed ideas is likely to frustrate and annoy any
assessor – not a good recipe for obtaining good marks.

My students performed well at school and are frustrated at losing marks over what seems
to them unimportant details. However, by the end of the year they generally accept that
writing well has improved their performance. You have to trust me that this works! Besides,
writing well in any subject is a useful skill to possess.

---

[1] To be honest, students don't mind this!

Writing well is good for you

Writing well

There are many reasons for writing – you might be making notes for future use or wish to communicate an idea to another person. Whatever the reason, writing mathematics is a difficult art and requires practice to produce clear and effective work.

Good writing is clearly important if you wish to be understood, but it has a bonus: it clarifies for you the material being communicated and thus adds to your understanding. In fact, I believe that if I can't explain an idea in writing then I don't understand it. This is one reason why writing well helps you to think like a mathematician.

Generally, we write to explain to another person, so have this person in mind. Two points to remember:

- Have mercy on the reader. Do not make it difficult for them – particularly someone marking your work.
- The responsibility of communication lies with you. If someone at your level can't understand it, then the problem is with your writing!

What follows is a collection of ideas on how to improve your writing. The ideas presented have been tried and tested with students over many years and are not merely theoretical ideas. They may seem troublesome and pedantic, but if you follow them you will produce clearer explanations, and hence gain more marks in assessments.

It should be noted that there is a huge difference between finding the answer to a problem and presenting it. These rules apply to the final polished product. When trying to solve a problem or do an exercise it is acceptable to break all these rules. What is important is that they are followed when writing up the solution for someone else to read.

An example

In a geometry course I stated the Cosine Rule.

> **Cosine Rule:** Suppose that a triangle has edges of length a, b and c with the angle opposite a equal to θ. Then,
>
> $$a^2 = b^2 + c^2 - 2bc\cos\theta.$$

If you have not met this before, then this is a good chance to 'Check the text' as described on page 18. Try drawing some pictures and trying some examples. More techniques for investigating such a statement will be found in Chapter 16.

The cosine rule is a useful result which can be regarded as a generalization of Pythagoras' Theorem when we take $\theta = \pi/2$. (Check the text!) During the geometry course I proved this formula in the case that θ was an acute angle and left the case of an obtuse angle as an exercise. Figure 3.1 shows one solution I received. We will refer to this as we proceed. As an exercise take a look at it and try to spot as many errors as possible. Does it make sense? Is it easy to read? Most importantly, is it right?

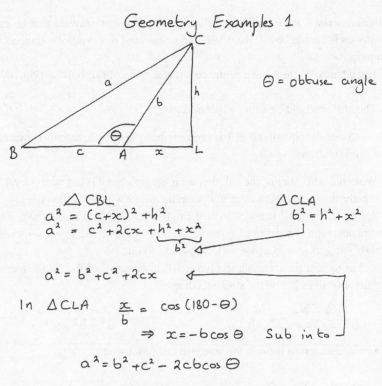

Figure 3.1 Student's proof of the Cosine Rule

Basic rules

The primary rule is that you should write in simple, correctly punctuated sentences. Let's put some more detail on this.

Write in sentences

Write in sentences. Write in sentences. And once more to really hammer it home: Write in sentences.

This advice has precedence over all others and is the one that can really change the way you present your work.

One of most common erroneous beliefs of the novice mathematician is that because mathematics is a highly symbolic language we need only provide a list of symbols to answer a problem. This is wrong, symbols are merely shorthand for certain concepts; they need to be incorporated into sentences for there to be any meaning.

Consider this student's answer to an exercise on finding the solution of a set of equations:

'$0 = 1$, $\therefore$ no solutions, empty set ($\emptyset$).'

It is obvious what the student meant: 'Since the equations reduce to the equation '$0 = 1$', which doesn't have any solutions, the solution set is empty.' This vital fact – that no solutions exist – is certainly included. The student also showed that he knows that the empty set is denoted by $\emptyset$. However, the inclusion of this symbol is unnecessary; it serves no purpose.

But what he wrote is not a sentence – it is a string of symbols and conveys no meaning in itself.

The answer could be better expressed as

> 'Since the equation $0 = 1$ is present, the system of equations is inconsistent and so no solutions exist.'

We could add 'That is, the solution set is empty', but it is not necessary. Understanding is clearly shown in this answer, and so more marks will be forthcoming.

All the other usual rules of written English apply, for example the use of paragraphs and punctuation. The rules of grammar are just as important: every sentence should have a verb, subjects should agree with verbs, and so on.

Let us look at the example in Figure 3.1 of the proof of the cosine formula. Examine the first two lines below the student's diagram.

$$\triangle\, CBL \qquad\qquad\qquad \triangle\, CLA$$
$$a^2 = (c+x)^2 + h^2 \qquad\qquad b^2 = h^2 + x^2$$

If I read from left to right in the standard fashion, I read

$$\triangle CBL\ \triangle CLA\ a^2 = (c + x)^2 + h^2\ b^2 = h^2 + x^2.$$

Now what does that mean? It is obvious what is intended. But why should we have to work out what was intended? It would be better to say what was meant from the start:

> In triangle $\triangle CBL$ we have $a^2 = (c+x)^2 + h^2$ and in $\triangle CLA$ we have $b^2 = h^2 + x^2$.

This is now a proper sentence. As an aside, notice how I explained my notation $\triangle$ by using the word 'triangle'.

Now look at the words after the $\Longrightarrow$ sign:

$$\frac{x}{b} = \cos(180 - \theta)$$
$$\Rightarrow\ x = -b\cos\theta \qquad \text{Sub into}$$

This is a perfect example of where we can understand what the student had intended but it is not well written. It is much clearer as

> $\ldots\, x = -b\cos\theta$. Substituting this into $\ldots$

Use punctuation

The purpose of punctuation is to make the sentence clear. Punctuation should be used in accordance with standard practice. In particular, all sentences begin with a capital letter and end with a full stop. The latter holds even if the sentence ends in a mathematical expression. For example,

'Let $x = y^4 + 2y^2$ Then x is positive.'

needs a full stop after the expression $y^4 + 2y^2$ as it is obvious that the second part is a new sentence – it begins with a capital letter. This is true for a list of equal expressions:

$$x = y^2 + 2y$$
$$= y(y + 2)$$

This should end with a full stop. Note that some authors do not adhere to this rule of punctuation. They are wrong.[2]

Mathematical expressions need to be punctuated. For example,

'Let $x = 4a + 3b$ where $a \in \mathbb{R}\ b \in \mathbb{Z}$'

should have commas like so

'Let $x = 4a + 3b$, where $a \in \mathbb{R}, b \in \mathbb{Z}$.'

Notice the three commas and the final full stop in the following example.

$$\text{Let } f(x) = \begin{cases} x^2, & \text{for } x \geq 0, \\ 0, & \text{for } x < 0. \end{cases}$$

Look at the example of the proof of the cosine formula. As you can see there is is no punctuation! Presumably a sentence starts at 'In $\triangle CLA$ ...' but it is not proceeded by a full stop so who knows?

Keep it simple

Mathematics is written in a very economical way. To achieve this, use short words and sentences. Short sentences are easy to read. To eliminate ambiguities avoid complicated sentences with lots of negations.

Consider the following hard-to-read example:

'The functions f and g are defined to be equal to the function defined on the set of non-positive integers given by x maps to its square and x maps to the negative of its square respectively.'

This would be better as:

'Let $\mathbb{Z}^{\leq 0} = \{\ldots, -5, -4, -3, -2, -1, 0\}$ be the set of non-positive integers. Let $f : \mathbb{Z}^{\leq 0} \to \mathbb{R}$ be given by $f(x) = x^2$ and $g : \mathbb{Z}^{\leq 0} \to \mathbb{R}$ be given by $g(x) = -x^2$.'

Note that we separated the definition of the domains of the maps into a separate sentence.

---

[2] A number of people think this is a controversial statement. 'What does it matter, as long as you are consistent?' Well, we could apply that argument to any sentence and we can get rid of all full stops! The majority opinion is that sentences end with a full stop – go with that.

Also we defined the set in words and clarified by writing it in a different way. The definitions of f and g are mixed together in the first sentence due to the use of 'respectively', while in the second sentence they are separated and defined using symbols. Sometimes using symbols is clearer, sometimes not; see page 28.

Expressing yourself clearly

The purpose of writing is communication – you are supposed to be transferring a thought to someone else (or yourself at a later date). Unfortunately – and I have lots of experience of this – it is easy to communicate an incorrect or unintended idea. The following advice is offered to prevent this from happening.

Explain what you are doing – keeping the reader informed

Readers are not psychic. It is crucial to explain what you are doing. To do this imagine that you are giving a running commentary. As stated earlier, it is not sufficient to produce a list of symbols, formulas, or unconnected statements. A good explanation will help gain marks as it demonstrates understanding.

You can introduce an argument by saying what you are about to do, e.g.

'We now show that X is a finite set',
'We shall prove that . . .'.

Similarly you can end by

'This concludes the proof that X is a finite set', or
'We have proved . . .'.

Make clear, bold assertions. Avoid phrases like 'it should be possible'; either it is possible or it isn't, so claim 'it is possible'. Be positive.

Of course, avoid going to the extreme of explaining every last detail. A balance, which will come from practice and having your written work criticized, needs to be struck.

If we look at the end of the example in Figure 3.1, then we see the following.

$$\Rightarrow \quad x = -b\cos\theta \quad \text{Sub into } \lrcorner$$
$$a^2 = b^2 + c^2 - 2cb\cos\theta$$

This ending would be better as

'... $x = -b\cos\theta$. Substituting this into the above we deduce that $a^2 = b^2 + c^2 - 2cb\cos\theta$.'

This is certainly much better as it implicitly makes the claim that what we had to prove has been proved. Otherwise it may look like we wrote the cosine formula at the end to fool the marker into thinking that the solution had been given. Also, using the word 'deduce' in the final sentence explains where the result came from.

Explain your assertions

Rather than merely make an assertion, say where it comes from. That is, use sentences containing

'as, because, since, due to, in view of, from, using, we have,' and so on.

For example,

'Using Theorem 4(i), we see that the solution set is non-empty'

is obviously preferable to

'The solution set is non-empty,'

and

'$x^3 > 0$ because x is positive'

is better than the bare

'$x^3 > 0$'

since, for a general $x \in \mathbb{R}$, we don't have $x^3 > 0$. The point is that the reader may be misled into thinking the statement is 'obviously false' if they had forgotten that x was positive. It doesn't hurt to include such helpful comments.

Another example is to say when a rule has been used:

'$f'(x) = 2x \cos(x^2)$ by the Chain Rule.'

In this way, you demonstrate your understanding.

Returning to the first few lines of the student's proof of the cosine formula in Figure 3.1

$$\triangle CBL$$
$$a^2 = (c+x)^2 + h^2$$

$$\triangle CLA$$
$$b^2 = h^2 + x^2$$

we have already seen that it would be better to have said

'In triangle $\triangle CBL$ we have $a^2 = (c+x)^2 + h^2$ and in $\triangle CLA$ we have $b^2 = h^2 + x^2$.'

But what about the next line? It says simply

$$a^2 = c^2 + 2cx + h^2 + x^2$$

Is this a deduction from the diagram? Certainly the first two equalities were, i.e. $a^2 = (c+x)^2 + h^2$ and $b^2 = h^2 + x^2$. In this case the line is not deduced from the diagram but from the first equation by expanding the bracket. So we should say so.

Expanding the brackets we get $a^2 = c^2 + 2cx + h^2 + x^2$.

We'll see that it is not necessary to phrase it this way when we look at the next line:

$$a^2 = b^2 + c^2 + 2cx$$

This comes from substituting the second equation, $b^2 = h^2 + x^2$, into the expanded version of the first, $a^2 = c^2 + 2cx + h^2 + x^2$. Let's say so.

> 'In triangle $\triangle CBL$ we have $a^2 = (c + x)^2 + h^2$ and in $\triangle CLA$ we have $b^2 = h^2 + x^2$. Expanding this first equation and substituting in b^2 from the second we get $a^2 = b^2 + c^2 + 2cx$.'

Note that we have left out the expansion of the brackets. You can include it if you wish but the calculation is so trivial that it is not worth the ink. The reader can check it themselves if they don't believe us.

Say what you mean

In any writing, saying what you mean is important – and difficult. Precise use of grammar can help in this task.

The first rule is that the reader should not have to deduce what you mean from context; all the necessary information should be there. Nothing should be ambiguous.

The true mathematician is pedantic, and requires that mathematics is precise. Without precision mathematics is nothing. Without it we cannot build with one concept placed on top of another. If one of the ideas is vague or open to different interpretations by different parties, then errors can creep in and the endeavour is unsound. So, be precise!

As an example, use the quantifiers 'some' and 'all'. Rather than say

$'f(x) = 5'$,

which is ambiguous – the reader may ask 'Is it for one x? At least one x? All x?' – say

$'f(x) = 5$ for some $x \in \mathbb{R}$', or $'f(x) = 5$ for all $x \in \mathbb{R}$',

depending on the situation.

More will be said in Chapter 10 on quantifiers to explain the importance of precision in this area.

Using symbols

We now come to tips concerning symbols. There is no escaping that mathematics is highly symbolic, but using lots of mathematical symbols does not make an argument a mathematical one.

Words or symbols?

Symbols are shorthand. For example, a famous theorem by Euler in the theory of complex numbers,

$$e^{2\pi\sqrt{-1}} = 1,$$

is concisely expressed in symbols.[3] The equivalent statement written out in words is less impressive:

'The exponential of two times the circumference of a circle divided by its diameter times the square root of minus one is equal to one.'

However, a good general rule of thumb is to use words. For example, use 'therefore' rather than the $\therefore$ symbol. Very few books use it. Similarly,

'x is a rational number $\Rightarrow x^2$ is real',

can be written as

'x is a rational number implies that x^2 is real.'

In some sentences it is best to avoid mixing symbols and words. For example,

'The answer $= 1$'

should be written as

'The answer equals 1.'

Otherwise we produce sentences like

'The number of people aged over $40 = 5$',

which reads all right, but the eye is drawn to the (erroneous) expression $40 = 5$.
    Small numbers used as adjectives should be spelled out, for example,

'the two sets'.

They should be in numerals when used as names or numbers, as in

'Lemma 3' and '… has mean equal to 23'.

Another example:

'One of the roots is 3.'

An exception is the number 1, which traditionally can be either.
    Note that symbols which are similar can cause confusion: clearly differentiate between $\in$ and ε. The former usually denotes membership of a set and the latter is the Greek letter epsilon, but be aware that other writers use them the other way round.
    As noted earlier we rewrote the first few lines and to include the standard notation $\triangle CBL$:

'In triangle $\triangle CBL$ we have $a^2 = (c+x)^2 + h^2$ and in $\triangle CLA$ we have $b^2 = h^2 + x^2$.'

---

[3] This is a great theorem – it relates many great numbers, e, π, the square root of -1 and of course two important natural numbers: 1 and the only even prime, 2. In a poll of mathematicians (*Mathematical Intelligencer*, Vol. 12 no. 3, 1990, pp. 37–41), this theorem was voted the most beautiful theorem in mathematics.

Equals means equals

The equals sign, =, is one of the most common in mathematics, and one of the earliest learned by children. Despite this, or maybe because of it, it is still badly abused.

Let's go back to the beginning and note that, in using the equals sign, we are asserting that the two objects on either side are *exactly the same* – being almost the same is not enough, being close is not enough, being similiar in a poor light and from a distance is not enough! For numbers this idea of equality should be second nature. But it holds for other objects. Thus remember:

Equals means equals.

One consequence is that if on one side of the sign there is a function, then on the other side there must be a function. If on one side there is a set, then on the other there must be a set.

In answer to a question on factorising numbers into primes, one of my students wrote:

Factors: $6 = 2$ and 3.

Leaving aside the observations that this is a poor sentence and '$6 = 2$' is not good on the eye, the idea expressed is false. True, 6 is *equal to the product* of 2 and 3, and so has 2 and 3 as factors, but it is not *equal* to '2 and 3'. A better answer is:

The prime factors of 6 are 2 and 3.

Similarly, consider the exercise, 'Find the derivative of x^3', the answer is not

$$x^3 = 3x^2.$$

Now, this does give a mathematical expression, in fact an equation, but it is not what the student wanted to assert. One correct way to write it is

$$\frac{\mathrm{d}}{\mathrm{d}x}\left(x^3\right) = 3x^2.$$

A very common mistake is to use the equals as a link from one line to the next, almost like a sign saying this is the next part of the process. The correct way of displaying results is given next.

Displaying results with the equals sign

If an expression is short, we show working by writing across the page. For example, $(x + 3)^2 = (x + 3)(x + 3) = x^2 + 6x + 9$.

For a longer calculation it is traditional to write down the page like so:[4]

$$\begin{aligned}
(x+3)^2 + x^2 &= (x+3)(x+3) + x^2 \\
&= x^2 + 6x + 9 + x^2 \\
&= 2x^2 + 6x + 9.
\end{aligned}$$

---

[4] Unfortunately, this violates our rule on punctuation, but we do it anyway as it is practical and traditional.

Sometimes we need to indicate where a particular result came from. Avoid interrupting the flow of the argument like so:

$$= x^2 + 5y$$
$$\text{by theorem } 6 = x^2 + 25 \ldots$$

If the details of why a particular step is true need to be included, then do the following. For the sake of argument suppose that $y = 3$ by Theorem 4.6. Then we write

$$x^2 + 4x + y = x^2 + 4x + 3, \text{ by Theorem 4.6,}$$
$$= (x + 1)(x + 3) \ldots$$

Note the punctuation after the symbolic expression on the first line and after the mention of the theorem. It doesn't read well, but is clear on the page.

Don't draw arrows everywhere

If a result requires an earlier one, it is tempting to draw a long arrow to point to it. Don't do this on aesthetic grounds. Instead, give the required result a name, number or symbol, so you can refer to it.

Our example in Figure 3.1 uses arrows.

We can change this to

'Expanding this first equation and substituting in b^2 from the second we get $a^2 = b^2 + c^2 + 2cx$. $(*)$

$$\vdots$$

$\ldots \implies x = -b\cos\theta$. Substituting this into $(*)$ we deduce that $a^2 = b^2 + c^2 - 2cb\cos\theta$.'

Exercise 3.1

Rewrite the proof of the the Cosine Rule so that it follows the suggestions given.

Finishing off

Proofread

Always proofread your work. That is, read through it looking for errors. These could be typographical errors (also known as typos), where the wrong character is used, e.g. cay instead of cat, or spelling mistakes, e.g. parrallel instead of parallel, grammatical mistakes, e.g. 'A herd of cows are in the field', or even mathematical errors.

Read your work slowly. Reading aloud can help catch many errors as it stops you skimming. Get someone else to read your work as you will often read what you think is there, rather than what actually is there. If your checker misses mistakes, then you are not allowed to blame them. The final responsibility always rests with the writer!

A useful proofreading method is to concentrate on one aspect of proofreading at a time. That is, read through first for accuracy, i.e. is it true? Next, check for spelling, typos, are all the brackets closed?, etc. After that check that the order of the material is correct and that it flows as you read it.

Reflection

Reflection is an important part of the writing process. Put your work away for some time and come back to it with a fresh eye. Obviously, this is not possible for work with tight deadlines, but can be done with project work.

When reading through again, ask 'What can I take away?' (aim for economy of words) and 'What can I add?' (more examples might clarify). For the former remove unnecessary words and sentences. Also ask: 'Are all the symbols explained and are they necessary? Does it say what I mean and is it simple? Is it more than just a collection of symbols?' And of course, most importantly, 'Did I write in sentences?'

Exercises

Exercises 3.2

(i) **The Sine Rule:** Suppose that we have a triangle with sides of length a, b and c with the angles opposite these sides labelled α, β and γ respectively. Then

$$\frac{\sin \alpha}{a} = \frac{\sin \beta}{b} = \frac{\sin \gamma}{c}.$$

In an exam a student answered the question 'State and prove the Sine Rule' with the following:

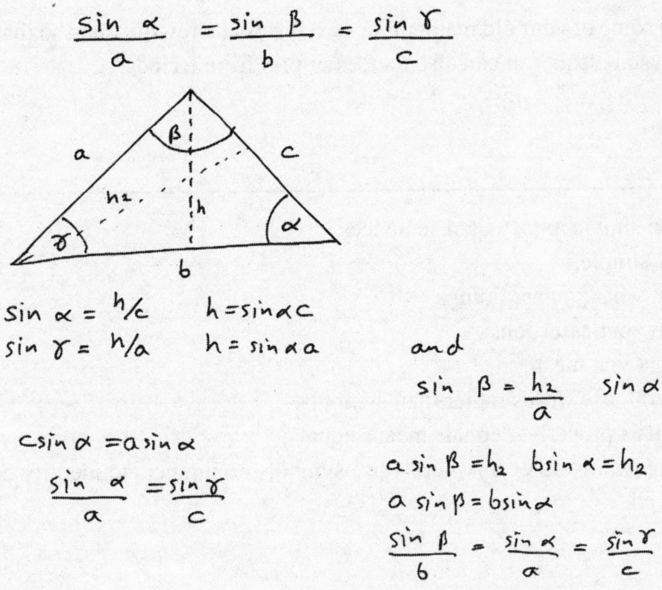

$$\frac{\sin \alpha}{a} = \frac{\sin \beta}{b} = \frac{\sin \gamma}{c}$$

$$\sin \alpha = \frac{h}{c} \qquad h = \sin \alpha \, c$$
$$\sin \gamma = \frac{h}{a} \qquad h = \sin \alpha \, a$$

and

$$\sin \beta = \frac{h_2}{a} \qquad \sin \alpha = \frac{h_2}{b}$$

$$c \sin \alpha = a \sin \alpha$$

$$\frac{\sin \alpha}{a} = \frac{\sin \gamma}{c}$$

$$a \sin \beta = h_2 \quad b \sin \alpha = h_2$$
$$a \sin \beta = b \sin \alpha$$
$$\frac{\sin \beta}{b} = \frac{\sin \alpha}{a} = \frac{\sin \gamma}{c}$$

Figure 3.2 Student's proof of the Sine Rule

Rewrite this answer so that it is correctly written and easily comprehended.

(ii) If you know how to find maxima and minima as well as curve sketching you should rewrite the following answer to the exercise 'Find the maximum and minimum values of the function $f(x) = 2x^3 - 12x^2 + 18x$ and sketch its graph.'

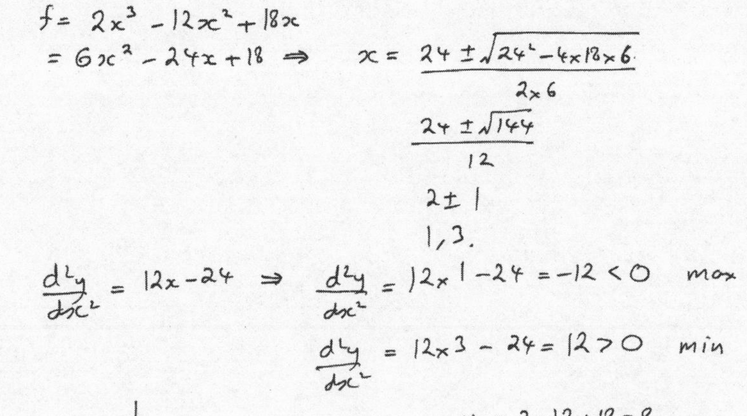

$$f = 2x^3 - 12x^2 + 18x$$
$$= 6x^2 - 24x + 18 \Rightarrow \quad x = \frac{24 \pm \sqrt{24^2 - 4 \times 18 \times 6}}{2 \times 6}$$
$$\frac{24 \pm \sqrt{144}}{12}$$
$$2 \pm 1$$
$$1, 3.$$

$$\frac{d^2y}{dx^2} = 12x - 24 \Rightarrow \quad \frac{d^2y}{dx^2} = 12 \times 1 - 24 = -12 < 0 \quad \text{max}$$

$$\frac{d^2y}{dx^2} = 12 \times 3 - 24 = 12 > 0 \quad \text{min}$$

$$y = 2 - 12 + 18 = 8$$
$$y = 2 \times 27 - 12 \times 9 + 18 \times 3$$
$$= 0.$$

Figure 3.3 Student's answer to finding maximum and minimum values of $f(x) = 2x^3 - 12x^2 + 18x$

(iii) Find some of your old mathematics exercises and rewrite them so that the exposition is crystal clear. You can also take examples from friends.

Summary

▶ Write in simple, punctuated sentences.
▶ Keep it simple.
▶ Explain what you are doing.
▶ Explain your assertions.
▶ Say what you mean.
▶ In general, use words rather than symbols.
▶ Use equals properly – equals means equals.
▶ Don't draw arrows everywhere – use symbols or numbers to identify equations.
▶ Proofread.
▶ Reflect.

CHAPTER 4

Writing mathematics II

Learn as much by writing as by reading.
Lord Acton, *Lectures on Modern History*, 1906

In the previous chapter we were concerned with the basic principles of writing mathematics. Here we shall become more specific and give particular examples of ways to improve the presentation of mathematics.

Expressing yourself clearly

If you use 'if', then use 'then'

If you use the word 'if', then use the word 'then' as well. (It is traditional to punctuate with a comma before the word 'then' but this seems to be dying out.) That is,

'If x is odd, then x^2 is odd'

is preferable to

'If x is odd, x^2 is odd.'

Statements can be written to include 'if' but for which 'then' is not needed:

'x^2 is odd if x is odd.'

But, in general, use 'then' as confusion may result when it is not employed. For example, what is the meaning of

'If $a > 0, b > 0, a + b > 0$'?

It could mean

'if $a > 0$, then $b > 0$ and $a + b > 0$.'

This is because maybe a is positive and so forces b to be positive, for example, $b = 5a$. Alternatively, it could mean

'If $a > 0$ and $b > 0$, then $a + b > 0$'

which is always true. The point is that omission of 'then' can lead to ambiguity. In many cases the reader can deduce the meaning of a statement with the 'then' omitted, but it is safer to include it. The reader should not have to deduce the meaning from other clues. Remember, the responsibility of communication rests with the writer.

Not everything is a 'formula': Call things by their correct name

A lot of attention is paid to formulas in mathematics and so many beginners call any collection of symbols a formula or an equation. Generally, such a collection is called an **expression** or **term**, e.g. $3x^2 - 7x$ is an expression.

An **equation** involves stating that two expressions are equal, for example, $3x^2 - 7x = 4x$. Note that an **inequality**, such as $x \leq 5$, is not an equation as an *equa*tion should be an *equa*lity.

A **formula** expresses some relationship or rule. It is often used when a method of calculating something from another expression is given, e.g. for studying $ax^2 + bx + c = 0$ we look at $D = \sqrt{b^2 - 4ac}$. This is the formula for D (sometimes known as the discriminant).

So, ensure that you call an object by its correct name. Care should also be taken to distinguish between concepts, for example, between being a set and being an element of a set.

Another example is that you should distinguish between a function and its value at a point, that is, between f and $f(x)$. The first represents a function which we have called f, while $f(x)$ is the *value* of the function f at x (see page 10). Strictly speaking, you should not call $f(x)$ a function, but this distinction is often ignored by mathematicians who should know better (and that includes me).

In summary, from the above, $f(x) = 3x^2$ is an equation (or expression or formula) which gives the values for a particular function f.

Avoid 'it'

Be explicit about what you are talking about. The word 'it' is a very useful word in the English language since it allows us to talk about something without really saying what that something is. Unfortunately, 'it' can be ambiguous. When writing mathematics, the word can be used as a way of disguising that we don't really know what we are talking about. Mostly this is self-deception – we do it subconsciously rather than deliberately.

If you find that you have used the word 'it', then replace it with the proper word or phrase. Does the sentence still make sense? If not, change it.

Decimal approximations

We tend not to use decimal approximations of numbers in pure mathematics but they are common in applied mathematics or statistics. This does not mean that they are never used in pure mathematics. If you are drawing figures, then it is permissible to use decimal approximations: for example, in finding where a quadratic crosses the axes.

So in pure mathematics if the final answer is $\sin 7$, then leave it as that rather than say 0.656986598. If you have to use approximations, then use the symbols $\simeq$ or $\approx$. So $\pi \approx 3.14$ and $\sqrt{2} \simeq 1.41$ are both acceptable. The point is that writing $\pi = 3.14$ is actually wrong – recall that equals means equals, page 30.

For expressions like $\sin(\pi/6)$ use $1/2$ or 0.5. Note that the use of 0.5 is all right as this is not an approximation - we are against approximations, not decimals.

Words or symbols?

Don't begin sentences with a symbol

Do not begin sentences with symbols. Thus, the next sentence should not be used.

'f is a function with domain $\mathbb{R}$.'

This avoids violating the rule that every sentence begins with a capital letter, but it applies even if the symbol is a capital, for example, 'X is a finite set' is bad. We could get a sentence like

'Suppose that x is an element of X. x is not in Y.'

Given that, in mathematics, the full stop functions as a multiplication symbol, then the sentence might be read as having X times x in the middle.

To avoid the problem use a description of the object the symbol represents. For instance, we can rewrite the earlier incorrect sentences as

'The function f has domain $\mathbb{R}$.'

'The set X is finite.'

and

'Suppose that x is an element of X and x is not in Y.'

Another approach is to employ phrases like 'We have' to begin sentences:

'We have $g(x) = 2x^4 - 5x - 3$.'

The curse of the implication symbol

A symbol commonly abused and overused by the novice mathematician is the symbol of implication: $\Rightarrow$. First, this should be read as 'implies' or 'implies that', and should not be used as an equals sign, so do not write $5 - 3 \Rightarrow 2$.

As we shall see in later chapters (and there is no harm in stating it now) the correct usage is

'statement $\Rightarrow$ statement.'

For example,

'x is odd $\implies x^2$ is odd.'

In much the same way that no sentence would begin with an equals sign, no sentence would begin 'implies that' so never begin one with $\Rightarrow$. If you feel the need to begin with an implication, then it is probably better to write 'This implies that ...'. Similarly many students use $\Rightarrow$ as a method of connecting one line to the next, something like 'this is what we do next'.

If you are unsure where to use it, then say the sentence out loud with the symbol $\Rightarrow$ replaced by 'implies' or 'implies that' as often this will help.

Similar problems are found with the symbol $\iff$, the biconditional; see Chapter 9. This is used to assert that two statements are equivalent. Strictly speaking, we should include them when writing a long list of equals signs. An example of this will be given in Chapter 21.

If you are unsure how to use $\implies$ or $\iff$, then consult Chapters 7 and 9.

Use common symbols and notation

Some symbols are given a common fixed meaning. The prime example of this is π, which represents the circumference of a circle divided by its diameter. It would be unusual to denote this ratio with anything other than π, but it is technically correct to start an argument with 'Let α be the circumference of a circle divided by its diameter.'

However, π can mean other things as well, e.g. it is often used for a projection map (since π represents p, and p is the first letter of projection). It is also used for the fundamental group of a space in algebraic topology.

Other symbols, such as ε, usually mean a small positive number, n is a natural number and so on. Get to know these conventions and use them as they make your work easier to read. See Appendix B for a list of examples.

Define your symbols and notation

As we have seen, many symbols get used for certain objects, such as f regularly denotes a function, and you may have noticed that I have been using capital letters such as X and Y for sets, and lower case letters such as x to denote the elements of those sets.

Despite these conventions you should define your notation so that it is totally unambiguous since a reader may use a different notation to you. Thus, write 'Let X be a set' rather than just use X without explanation. Another example of this can be seen in Exercise 3.2 where the student has given a statement of the Sine Rule where the letters a, b, c, α, β, γ have not been defined. Did you include a definition of the letters in *your* answer to that question?

Some notation does not need introduction; most mathematicians will understand what π stands for, provided it is in context, and will know that Σ refers to summation, $\int$ to integration and so on.

Making improvements

Use connecting phrases

Another common problem in writing is that assumptions and deductions in an argument are not clearly distinguished. To avoid this problem you should, in constructing an explanation, use connecting words and phrases, such as

'hence, as, therefore, since, and so,'

to indicate that implications and deductions are being made. If you look at a mathematics book, then you will see that there are plenty of uses of since, hence and therefore.

As an example, consider explaining that the order of brackets is important when dividing numbers. It is not sufficient to write

'$(8/2)/4 = 1, 8/(2/4) = 16$.'

We should say

'We have $(8/2)/4 = 1$, but $8/(2/4) = 16$.'

Notice that the 'but' brings attention to the important idea that there is a difference between the calculations. Using the word 'and' would not make the statement incorrect, but attention would not have been drawn.

Use synonyms

Repetition of words can lead to boredom for the reader. The use of synonyms helps to make the material more interesting.

Some synonyms for deduction are:

'hence, so, it follows, it follows that, as a result, consequently, therefore, thus, accordingly, then'.

Like most synonyms there can be slight differences in usages; one cannot always replace 'hence' by 'then'.

Synonyms for explanations are:

'as, because, since, due to, in view of, owing to'.

We can use a construction involving 'let' in place of one involving 'suppose'. For example,

'Let X be a set'

in place of

'Suppose that X is a set.'

Note that we say 'suppose that', i.e. 'suppose' is followed by 'that'.

Exercises

Exercise 4.1

(i)   Can you improve on your rewriting of the mathematics in the exercises of the previous chapter?

(ii) Now, more generally, find some of your previous answers to exercises and rewrite them using the guidance from this chapter.

Summary

▶ If you use if, then use then.

▶ Not everything is a formula. Call things by their correct name.

▶ Don't use the word 'it'.

▶ Avoid decimal approximations in pure mathematics.

▶ Don't begin sentences with a symbol.

▶ Use the implication symbol, $\implies$, correctly.

▶ Use common symbols and notation and define them first.

▶ Use connecting phrases and synonyms.

CHAPTER 5

How to solve problems

It isn't that they can't see the solution. It is that they can't see the problem.
G.K. Chesterton, *The Scandal of Father Brown* – 'The Point of a Pin'

Solving mathematical problems is hard and there is no magic formula which solves all problems. If there were, anyone could do it, and employers would not be so keen to hire problem-solving mathematicians. However, in this chapter we will try to provide at least some ideas on how to solve problems.

Let us first distinguish between an exercise and a problem. An exercise is something that can be solved by a routine method, for example finding the roots of a quadratic. A problem is something that will require more thought, for example, we have to draw together a number of ideas and perhaps apply a number of the routine methods learned through exercises, such as root finding, in a new combination to get the answer. However, in this book, to save writing out the phrase 'exercises and problems', we will group problems in with exercises. Nonetheless, knowing that there is really a distinction can help.

Much lower-level mathematics involves selecting and applying the right technique or formula to answer a question. While this is also true for some higher-level mathematics, we have the task of verifying the truth of statements, e.g. 'There is an infinite number of prime numbers.' The techniques needed for this type of problem are different from those of an apply-the-right-technique exercise.

Numerous books have been written on problem-solving. Reading one or two may be worthwhile, but the best way to learn how to solve problems is to solve problems. Experience counts. Despite saying this, there are useful tips which you can apply in your problem-solving. In this chapter I will present Polya's four-point plan for problem-solving and some advice on how to apply it in mathematics. You should first consciously apply the plan and techniques and then slowly over time push them into your subconscious.

Sample problems

To discuss problem-solving we need some problems to solve. Here are three.

(i)   We can define the **factorial** of a natural number n by multiplying all numbers up to and including n. We denote factorial n by $n!$ and read it as 'n factorial'. For example, $5! = 5 \times 4 \times 3 \times 2 \times 1 = 120$.

Problem: How many zeroes are at the end of 100!.

(ii)   Recall the definition of the cardinality of a finite set from page 6. It is just the number of elements in that set.

Problem: Suppose that X and Y are two finite sets. Find a formula that relates $|X|$, $|Y|$, $|X \cup Y|$ and $|X \cap Y|$.

(iii) Show that the equation $x^2 + y^2 = z^n$ has positive integer solutions for every $n = 1, 2, 3, \ldots$.

Exercise 5.1

Attempt to solve these problems before reading the following. Think about how you approach the problems and what techniques you are using. At this stage I do not expect you to find an answer; the problems are set to get you thinking and to think about thinking!

Problem (i) is a classic problem. It shows why we need mathematical theory. You *could* attempt to calculate 100! by hand but it would become very hard very soon. By developing a theory we not only find the answer but could also generalize our reasoning to calculate the number of zeroes at the end of 1000!, 10 000!, or even 99! or 97!.

Polya's four-step plan

The classic book on problem-solving is *How to Solve It* by G. Polya. For solving any type of problem – not just mathematical ones – he gives a four-point plan:

(i)   Understand the problem.
(ii)   Devise a plan.
(iii) Execute the plan.
(iv)   Look back.

At first glance this seems a trivial plan – parts (ii) and (iii) look suspiciously like 'solve the problem' – yet it does often provide insight. In the following the hints are grouped under the headings of this plan. These hints are intended to help, not to provide a magic formula for problem-solving. There is no magic formula. The hints do not have to be executed in order; jumping backwards and forwards between them may be required.

Understanding the problem

First we need to understand and explore the problem. Through experimentation and exploration convince yourself that a particular statement is true.

Understand all the words and symbols in the problem

A common problem that I face when dealing with students who are stuck on a problem is that often they don't understand all the words and symbols. When I ask them what a certain word means they are unsure. Take problem (ii) above; if you don't know what cardinality means or what the symbols | | mean, then there is no hope of answering the question.

To some extent this simple mistake by students is caused by their earlier experiences of mathematics. High-level mathematics involves understanding concepts much more than lower-level mathematics does. The latter by its nature involves problems where some well-defined procedure can be followed, e.g. finding a derivative by the chain rule. In these problems students can get very far by using a worked example as a template and they don't even have to really understand the definition of derivative.

At a higher level it has to be realized that the key to a solution may be to use the conditions of some particular definition. If you don't know that the definition provides that condition, then you are unlikely to solve the problem. So ensure you *know* and *understand* all the necessary definitions.

Guess – use your intuition

Guessing is good. Or rather, educated guesses are good – wild and unsubstantiated guesses are not.

You can develop your intuition by making an educated guess to the solution of a problem. This may only be a guess as to the method needed or some range of values within which a solution lies but is worth trying.

Exercise 5.2

What is an educated guess for each of the three problems above?

What do you know about the hypothesis and conclusion?

A problem can be divided into two parts: the **hypothesis** and the **conclusion**. The hypothesis is what we are given in the problem, i.e. what we know: for example, x, y and z satisfy the equation $x^2 + y^2 = z^n$; or the sets A and B are finite. Since the hypothesis often contains a number of parts we sometimes refer to the hypotheses of the problem. The hypothesis is also known as the **assumption** (or assumptions).

The conclusion is what we want to know or what we want to produce. For example, that we can find x, y and z that are integers, or find a formula relating the cardinality of A, B, $A \cup B$ and $A \cap B$.

A good strategy to solve problems is to write down what you know about the hypothesis and the conclusion of the theorem. That is, what do you know and what do you want to know. This may help trigger ideas.

Work backwards and forwards

In finding the solution, you do not have to begin with the assumptions and work towards the conclusion. You can start with the conclusion and try to think what this would imply. In this way one can work towards the assumptions. Once the solution becomes clear rewrite with the assumptions going to the conclusion. For example, if 100! has exactly x zeroes at the end what does that imply? It means that $100! = y \times 10^x$ for some y where y does not end in a zero. Thus, from 100! we need to take out factors of 10. Now we look at 100! and see from the definition how factors of 10 might arise. Proceeding in this way, i.e. backwards and forwards, we can produce an argument that 'meets in the middle'.

A danger with this method is that we assume what is to be proved. This problem will be investigated in more detail in Chapter 21.

Work with initial and special cases

Many problems will have some form of index. For example in problem (iii) the index is n. It can be any natural number and we have to find integers x, y and z for every possible n.

In these indexed problems you should try to solve the problem in the initial cases, e.g. $n = 1, 2$ and 3. This will not necessarily give you the general answer but allows insight and a 'feel' for the problem.

In problem (iii) we see that if $n = 1$ we have to find solutions to $x^2 + y^2 = z$. This is easy, take any integers x and y, then $x^2 + y^2$ is an integer; let's take it to be z. For example, let $x = 3$ and $y = 5$, then $x^2 + y^2 = 3^2 + 5^2 = 9 + 25 = 34$. Thus $(x, y, z) = (3, 5, 34)$ is a possible solution. Recall that the question said show that solutions exist for all n; we don't have to find *all* possible solutions for a particular n.

For $n = 2$ we have to solve $x^2 + y^2 = z^2$. This is the famous Pythagoras equation, for which we have many solutions, e.g. $3^2 + 4^2 = 5^2$.

For $n = 3$ we have to solve $x^2 + y^2 = z^3$. Play with this equation and see if you can use the $n = 2$ case to provide a solution. I am not suggesting that this will give you an answer; I am suggesting it because it is what I would try, i.e. use the answer in one case to find the answer in another. It may or may not work; the point is to try. Can you think of a different method of attack?

Work with a concrete case

In a similar way to the above idea, for an abstract problem look at a concrete case. In a problem concerned with sets take a specific set. So for problem (ii) take a set where $X = \{a, b, c, d\}$ and $Y = \{b, d, e, f, g\}$. In this case we see that $|X| = 4$, $|Y| = 5$. We also see that $|X \cup Y| = |\{a, b, c, d, e, f, g\}| = 7$ and $|X \cap Y| = |\{b, d\}| = 2$. So we need a formula that relates 4, 5, 2 and 7. Play around with some other examples if you don't see a pattern. Vary X and Y by an element or two and see how the cardinalities change.

Be aware though that when proving a general theorem it is not enough to show that it holds in one or two specific cases. More will be said on this in later chapters. The examination of specific cases is intended to provide insight and frequently unlocks the problem.

Draw a picture

The human mind is very good at working with images. Pictures are excellent for developing intuition about a problem and subsequently suggesting a solution to it. In fact, a diagram is often essential in problems from geometry or physics.

In the case of problem (ii) drawing a Venn diagram is very illuminating.

Think about a similar problem

As stated earlier, the way to become good at solving problems is to solve problems. Similar problems often have similar solutions, so consider problems with similar assumptions or conclusions that you have solved before and see if the same method will work.

In the $n = 2$ case of problem (iii) there is a way of constructing an infinite number of solutions; these solutions are called *Pythagorean triples*. It may be worth investigating these to see if the method of finding Pythagorean triples can be adapted to the solution of our problem. (Again, I am not saying it can be, I am saying it is worth thinking about!)

Find an equivalent problem

Reformulating the problem can be helpful. For instance, suppose that the problem was to show that two functions were equal. This is equivalent to showing that their difference is zero. We can say this more precisely in the following. Suppose that f and g are functions of a real variable. Define a new function h by

$$h(x) = f(x) - g(x).$$

Then

'$f(x) = g(x)$ for all x' is equivalent to '$h(x) = 0$ for all x'.

Now look for theorems with the conclusion that a function is the zero function – maybe that will give the answer.

Solve an easier problem

This is similar to working with a special case. With the experience from solving an easier problem it might be possible to solve the general one.

So for example, in problem (i) it may be easier to work with 10! rather than 100!. Try it and see what happens. It may suggest some ideas. In problem (iii) consider what happens if z is odd or even.

Rewrite in symbols or words

Mathematics can be described in words and in symbols. It is often useful to swap between the two. A good example can be found in Example 24.8. We will look at the idea of words versus symbols again in Chapter 16.

Devising a plan

Once the problem is understood construct a plan to solve it.

Break the problem into pieces

One of the first approaches is to break the problem into pieces; the hope is that each piece is an easier problem, preferably a mere exercise. Hence, divide the problem into as many natural cases as is possible.

For example, when differentiating a function we rarely see examples where the function is a simple product, say $x^2 \cos(x)$. We see much more complicated examples like $x^3 \cos(x^2 + 5) + e^{\cos(2x)}$ where we have to use a number of different rules, product rule, chain rule, etc. So we break the problem down into different pieces according to these rules.

Now consider problem (iii); it involves natural numbers. The set of natural numbers can be divided into two: even numbers and odd numbers. Let's look at the case of even numbers, that is, $n = 2m$ for some natural number m. (If we looked at n odd, then we would take $n = 2m + 1$ for some m.)

The equation becomes:

$$x^2 + y^2 = z^{2m} = (z^m)^2.$$

Thus we have something of the form $x^2 + y^2 = w^2$ (where $w = z^m$). This looks like the case where $n = 2$. Maybe we can use the solution we got in that case to help us.

Later we shall see that, to show two sets A and B are equal, a good strategy is to show separately that $A \subseteq B$ and $B \subseteq A$. Showing that one set is contained in another is an easier task.

The point to stress here is that if we can't answer the whole problem, then maybe we can answer part of the problem. Finding a partial answer often gives a feeling of accomplishment and hence encouragement that the problem can be defeated.

Find the right level

In mathematics problems there are levels at which one can work. Some problems can be solved by applying a theorem, others by applying a definition. We shall see plenty of examples of this in Parts IV and V.

Give things names

Giving a name to an object helps get a handle on it and allows one to translate a problem into mathematical language. (This will follow naturally from the process of changing from words to symbols.) Once in this language we have a hope of solving the problem.

For example, suppose the problem includes the information that a tank of water is emptying at a rate of $10 \, \text{m}^3$ per hour. In this case we would give the volume of water the letter V (for volume, of course), and then we can record the rate of volume change as $dV/dt = -10$ (it is negative as the volume is decreasing).

Systematically choose a method

It is helpful to know the standard methods and to apply them systematically and *consciously*. When integrating a function, methods such as integration by parts and substitution are available. Rather than staring at the question waiting for the correct method to make itself known, be systematic and ask 'Can I solve this by integration by parts? What would be the parts?' If that method doesn't work, ask 'Can I solve it by substitution? What sort of substitutions might work?' and so on.

The same is true of proving a statement. As we shall see in Part IV, there are many methods of proof available to the mathematician, for example the direct, contradiction, contrapositive, induction and exhaustion methods. However, as it is the most obvious – one goes directly from the assumptions to the conclusions – the direct method is the one usually chosen by the novice. Deliberately trying the various other methods in turn and seeing whether they provide the answer can be productive. More will be said on these methods later.

Executing the plan

Once the plan has been made, it needs to be carried out. By now you should have a good idea of why a result is true, next work out precisely what convinced you and polish that.

Check each step – don't use intuition

Check each step carefully and ensure that it is justified. Any argument is only as strong as its weakest link. If only one small step is incorrect, then the whole argument is false. Now is the time to avoid using intuition. If you think something is true, then prove it beyond doubt by using small steps of logic. We shall see how to do this in Parts II and IV.

Looking back

Even though you have produced a solution, the problem-solving plan is not finished. As always, you should reflect on what you have done. Ultimately, what have you learned from solving the problem?

Check the answer

Verifying an answer is important. First, does the answer make sense – is it of the right order? If you have calculated that your car is travelling at a million miles an hour, or that the Sun is 299 km from the Earth, then it is unlikely that your answer is correct.

Lots of other tests are available. For example, the angles in a triangle should add up to 180 degrees. Another simple test involves the idea of **parity**. This is easier to see in action than give a good definition. The equation $51^2 = 34^2 + 46^2$ is false. This is easy to see without any calculation: The left-hand side is odd and the right-hand side is even. That's parity!

When solving equations a solution may be difficult to find but it is almost trivial to put that solution back into the equations to check that it really is a solution. For example, suppose we calculate that $x = -3$ is a solution of $2x^2 - x - 10 = 0$. We can check this easily. Put the x into the left-hand side and see if zero comes out:

$$2x^2 - x - 10 = 2(-3)^2 - (-3) - 10 = 2(9) + 3 - 10 = 18 + 3 - 10 = 11 \neq 0.$$

Thus, our calculation was wrong, $x = -3$ is not a solution. We can make similar checks for solutions of differential equations.

Similarly, if we integrate a function f to get F, then differentiating F must give f. This provides a simple check of integrals. It is worth doing.

One interesting by-product of this type of checking is that it sharpens our intuition without us realizing.

Find another solution

Even though a solution has been found there may be a better one, so attempt to solve the problem in a different way. Doing this may produce a better solution or identify a flaw in the present one.

Compare yours with model solutions, if they are available.

Reflect

Reflection really pays in solving problems. Think about what solved the problem and ask questions. How is the solution similar to others? How is it different? Was a certain theorem or technique used and does it keep getting used? Think about what keeps cropping up and put it in your armoury, ready to attack future problems.

Think about how the solution could be improved. Ask: What can be combined? What can be simplified?

Finally, try to show that your answer is wrong. This may sound like a strange suggestion but it can help convince you that your answer is correct and can find hidden errors.

Exercises

Exercises 5.3

(i)   Show that $\dfrac{a+b}{2} \geq \sqrt{ab}$ for $0 < a \leq b$.

(ii) Show that

$$a^2 + b^2 + c^2 \geq ab + bc + ca$$

for all positive integers a, b and c.

(iii) Let $f(x) = 1/(1-x)$. Define the function f^r to be

$$f^r(x) = \underbrace{f(f(f(\ldots f(f(x)))))}_{r \text{ times}}.$$

Find $f^{653}(56)$.

(iv) Show that

(a) $\sqrt[7]{7!} < \sqrt[8]{8!}$, and

(b) $\sqrt{100001} - \sqrt{100000} < \dfrac{1}{2\sqrt{100000}}$.

Hint in both cases: Try to get rid of square roots (via different methods).

(v) Suppose that three friends have a meal for \$25. They misread the bill, thinking it says \$27, give the waiter \$10 each and ask for a total of \$3 in change, i.e. \$2 less than they should get. The waiter puts \$25 in the till, gives each of the friends \$1 and secretly pockets the extra \$2.

Later the waiter does some calculations. The friends paid \$$(10 - 1) = 9 each which is \$27 and he kept \$2, giving a total of \$29. But the friends gave \$30. What happened to the extra \$1?

(vi) Bottle A contains a litre of milk and bottle B contains a litre of coffee. A spoonful of coffee from B is poured into A and the contents are mixed well. Liquid from A is then poured into B until B has one litre of liquid. Is the fraction of coffee in A greater the fraction of milk in B or is it the other way round?

Summary

- ▶ Use Polya's four-point plan.
- ▶ Understand all the words in the problem.
- ▶ Guess.
- ▶ Write down what you know about the hypothesis and conclusion.
- ▶ Work backwards and forwards.
- ▶ Work with initial, special and concrete cases.
- ▶ Draw a picture.
- ▶ Think about a similar problem.
- ▶ Find an equivalent problem.
- ▶ Solve an easier problem.
- ▶ Rewrite in symbols or words.
- ▶ Break the problem into pieces.
- ▶ Find the right level.
- ▶ Give things names.
- ▶ Systematically choose a method.
- ▶ Check each step.
- ▶ Check the answer.
- ▶ Find another solution.
- ▶ Reflect.

PART II

How to think logically

CHAPTER 6

Making a statement

When dealing with people, let us remember we are not dealing with creatures of logic.
We are dealing with creatures of emotion, creatures bristling with prejudices and
motivated by pride and vanity.
Dale Carnegie, *How to Win Friends and Influence People*, 1936

When I tell people that I am a mathematician it is often not very long before they bring up the subject of how logical mathematics is. Sometimes they consider this a downside to the subject as it makes it so much harder. After all, with logic you have to be right whereas in other areas of life opinions matter more. And it's easy to have an opinion – even if you can't back it up with evidence.

Logic is in essence quite simple despite its reputation for difficulty. A small number of simple rules exist. We need only to apply these and reduce complicated statements with them to achieve success.

In the next few chapters we shall concentrate on the logic used by mathematicians in their day-to-day work rather than on deep conundrums and paradoxes or on technical material such as predicates, compound statements and so on. I will, however, explain truth tables, which are rarely used by mathematicians in their everyday work, as they do provide a lot of clarity for beginners.

Mathematics is the business of proving mathematical statements to be true or false. So, first let's look at statements.

Statements

Defining precisely what is meant by a mathematical statement is surprisingly difficult – we could get into some very deep philosophical work at this point. However, I wish to be very practical – I want to give you the tools that a mathematician uses on a day-to-day basis – and hence we use the following definition.

Definition 6.1

*A **statement** is a sentence that is either true or false – but not both.*

This definition will be sufficient for our purposes. Some examples:

(i)   '$2+2=4$.' This is true.
(ii)   'All cats are grey.' A false statement, as some cats are black.
(iii)   'There are infinitely many primes.' True, to be proved in Chapter 23.
(iv)   '$0 = 1$.' Perfectly fine as a statement – it just happens to be false.
(v)   'Every integer greater than 4 is the sum of two prime numbers.' (Recall that a **prime number** is a natural number greater than 1 which can only be divided by itself and 1 without leaving a remainder.) This is called the **Goldbach Conjecture**. No-one knows whether it is true or not, but it is still a statement because it is either true or false.
(vi)   'Suppose $x^2 = 2$. Then x is not a rational number.' This will be proved in Chapter 23.

Not every sentence can be a statement as the following two sentences show.

(i)   'Open the window!'
(ii)   'x is an odd number.'

The first is a command and is fairly obviously not a statement.

The sentence

   'x is an odd number'

depends on x and so we cannot decide if it is true or false without further information. If x is 3, then the sentence is true. But if $x = 50$, then sentence is false. We need to know something more about x. Because there is a condition attached to the x we call this a **conditional statement**.

Technically speaking a conditional statement is not a statement according to the definition we gave above. However, in the interests of simplicity we shall refer to them as statements. (Just pretend that we have someone who can supply us with the extra information we need about x!)

Be aware that there are many sentences that are statements but we can never decide whether they are true or false. For example,

   'Last year I ate a total of 20 cheesecakes.'

This is a statement as the sentence is certainly true or false, but there is no way (without the aid of a time machine) we could decide.

An important part of the definition is that statements are either true or false – there is no in-between. This is known in technical language as **the law of the excluded middle**. That is, the idea of something lying in the middle between true and false is excluded from being a possibility.

An important example

A more subtle example of a sentence that is not a statement is

   'This statement is false.'

It looks like a statement and even claims it is one. Is it true or is it false?

First, let us assume that the sentence is true. Then the sentence says that the sentence is false. This contradicts our assumption that it is true. So, the sentence can't be true.

Now, assume the sentence is false. Then the sentence is correct, the statement is false. In other words, what the sentence says is true. But this contradicts the assumption that it is false.

Thus, we see that, whatever we assume, we get something that contradicts the assumption. Since we cannot sensibly decide whether the sentence is true or false it is not a statement. (The problem arises from the sentence being self-referential, i.e. it refers to itself. Strange things can happen in such sentences! Refer to a book on logic for a detailed discussion.)

To have a more mathematical feel we use letters to denote statements. Thus we talk of 'statement A' and so on.

Exercises 6.2

Which of the following are statements? Explain.

(i)   Aristotle was Greek.
(ii)   Aristotle was great.
(iii)   The number $\sqrt{2}$ is rational.
(iv)   The square root of an integer is a rational number.
(v) $3x^2 + 20x - 5 = 0$.
(vi)   Let x be an integer. Then $\sqrt{x}$ is rational.
(vii)   There is a number x such that $\sin(x) = x$.
(viii) There is an infinite number of natural numbers.
(ix)   There is an infinite number of rational numbers.

Using non-mathematical examples

Logic is an abstraction and so can be hard to grasp. To make it easier I often use non-mathematical statements or statements from everyday situations. For example, let us start with the fact that Winston Churchill (the British Prime Minister during most of the Second World War) was born in Blenheim Palace, England. From this we will take as a basic fact

'Winston Churchill is English.'[1]

We can also have

'President George Washington was the first President of the United States.'

We can use these, as we shall see later, to help clarify subtle points of logic.

Talking of subtle, try the following exercise from everyday language to test your powers of pedantry.

---

[1] It can be argued that the country of your birth does not determine your nationality. Let's keep things simple and avoid such arguments.

Figure 6.1 Exactly three coins on a table

Exercise 6.3

Imagine that there are coins on a table; you count them and find there are exactly three of them as in Figure 6.1.

Which of the statements are true?

(a) There are four coins on the table.
(b) There are two coins on the table.
(c) There are three coins on the table.
(d) There is a coin on the table.

Thought about it? Well, all of them are true except the first. Statement (a) is obviously false and Statement (c) is obviously true. Statements (b) and (d) can cause problems. Consider Statement (b). We know that there are three coins on the table because we counted them. We must have two there, and one other. But certainly we have two. Similar reasoning shows that Statement (d) is true as well.

The point is that the statement 'There are two coins on the table' is true but it is misleading because it contradicts everyday usage. If you asked 'How many coins are on the table' and I said 'two', then I think everyone would agree that I had misled you. To avoid this we use phrases such as 'There are exactly three coins on the table.' You need to be aware that mathematics requires pedantry. Become very precise in how you read statements.

Negation

Definition 6.4

*The **negation** of statement A is the statement that is false when A is true.*

For example the negation of

'Winston Churchill is English'

is

'Winston Churchill is not English.'

The negation of 'x is odd' is 'x is not odd'. (This can be re-expressed as 'x is even', of course.)

The negation of statement A is written as 'not(A)'. So 'not (x is odd)' is 'x is not odd' as above. Technical books on logic often use $\neg A$ instead of not(A).

Warnings!

Note that the negation of 'All cats are grey' is not 'All cats are not grey'. Recall, from our definition of negation, that not(A) is false when A is true. For the statement 'All cats are grey' to be false we need only have *one* non-grey cat, for example, a white one. So we want 'There exists a cat that is not grey' to be true. More will be said on such examples in Chapter 10.

Statement 'not(not(A))' is the same as statement A. That is, a double negative makes a positive. In some languages a double negative is used to emphasize a statement and does not make a positive. In colloquial English this is sometimes the case, such as in 'We don't need no education.' The speaker's intention is that no education is needed but in fact they are saying that an education is needed.

Truth tables

Truth tables are extremely useful when learning logic. Admittedly, mathematicians don't use them often in day-to-day work but they can provide clarity for beginners. The idea is to summarize in a table all possibilities for the truth or otherwise of a statement.

The truth table for 'not' is particularly simple:

A	not(A)
T	F
F	T

Here T means True and F means False. The information is read along the rows. So on the second line we see that if Statement A is True (given by T), then not(A) is False (given by F). On the third line we can see that A is false means that not(A) is true.

We can view the first column as the input and the second column as the output. If we put in F, we get out a T.

Example 6.5

Let us see the truth table for not(not(A)).

A	not(A)	not(not(A))		A	not(A)	not(not(A))
T	F			T	F	T
F	T			F	T	F

We begin by drawing the left-hand table. We then fill it in to produce the one on the right. We do this by going across the rows. When 'not(A)' is false, then we know from the truth table for 'not' that 'not(not(A))' is true. So, at the end of the second row, we enter a T. Similarly for the third row we fill in an F.

Since the first and third columns are the same we deduce that the statements 'A' and 'not(not(A))' are equivalent, i.e. essentially they are the same statement.

'And' and 'or'

We can build up statements using the English words 'and' and 'or'. Since they connect statements they are called **connectives**.[2]

Statements using 'and'

The 'and' connective is easiest as it corresponds to the standard English usage so let's do it first. Consider the example

'My dog is black and Paul's dog is white.'

This is a true statement when I have a black dog *and* Paul has a white one. It is false in all other cases. So if I have a white dog, then it doesn't matter what the colour of Paul's dog is, the statement is false. Similarly, if Paul has a brown dog and my dog is black, the statement is false.

To construct a truth table for 'and' we note that there are two inputs, A and B, the colours of the dogs (for 'not' there was only one input: A). Since each input can, independently, be true or be false we have *four* possibilities: Both false, A false and B true, A true and B false, and finally, both true. The truth table for 'and' using statements A and B is the following:

A	B	A and B
F	F	F
F	T	F
T	F	F
T	T	T

As you can see if just one of A or B is false, then the statement 'A and B' is false. We need both to be true.

Here the first two columns are the inputs and the last column is the output. Also, note that if we had three inputs, then we have $2^3 = 8$ different possibilities for the inputs being true or false.

Remark 6.6

Logicians use $\wedge$ to denote 'and', i.e. the statement 'A and B' is written as $A \wedge B$.

Now we can use truth tables to analyse complicated expressions. For example, when is 'A and not(B)' true? We can take the component parts A, B, and not(B), and build them up into the statement we want.

A	B	not(B)	A and not(B)
F	F	T	F
F	T	F	F
T	F	T	T
T	T	F	F

---

[2] Logicians go further and give the fancy names 'conjunction' and 'disjunction' to 'and' and 'or' respectively.

Note that we create column 3, not(B), by using the truth table for 'not'. We then create the final column using the truth table for 'and' applied to 'A' and 'not(B)', i.e. columns 1 and 3.

Statements using 'or'

In contrast to 'and', the use of 'or' is slightly different to standard English usage. For example, the statement

'David or Paul is going to the shop'

will usually mean that only one of them is going to the shop – not both. This type of 'or' is called **exclusive or**. The idea is that one part being true excludes the other being true.

In mathematics the above statement would mean at least one is going – maybe both. This is called **inclusive or**.

In everyday language exclusive or and inclusive or are generally not distinguished; one has to guess from context. In mathematics we use the inclusive form of or, i.e. in the above we know at least one is going to the shops. Consider the sentence

'Assume one of m or n is odd.'

For a mathematician this does not say that one is odd and the other isn't! It says one of them is odd and the other can be even or odd – we don't care which. This is unlike the usage in English so be careful.

Even more confusingly, we can use 'or' in an exclusive sentence. For example:

'Let x be a natural number. Then x is even or x is odd.'

In this case the 'or' is effectively exclusive because the property of being odd or even is exclusive, i.e. a number cannot be both at the same time.

Now let's consider the connective used in the dogs example.

'My dog is black or Paul's dog is white.'

This statement will be true if my dog is black. It will be true if Paul's dog is white. And it will be true if my dog is black and Paul's is white. The statement will be false if my dog is not black *and* Paul's dog is not white.

The truth table for 'or' should help clarify:

A	B	A or B
F	F	F
F	T	T
T	F	T
T	T	T

From the table we see that 'A or B' is false only when both A and B are false.

Remark 6.7

Logicians use $\vee$ to denote 'or'. That is, the statement 'A or B' is written as $A \vee B$.

Equivalence of statements

Now that we have complicated statements we can define what we mean by being the same in a mathematical way.

Definition 6.8

*Two statements are **equivalent statements** if the truth tables made from their inputs and outputs are the same.*

Examples 6.9

(i) The truth table for the statement A and the statement $\text{not}(\text{not}(A))$ can be summarized as

A	A
T	T
F	F

and

A	$\text{not}(\text{not}(A))$
T	T
F	F

.

Hence A and $\text{not}(\text{not}(A))$ are equivalent statements.

(ii) The statements 'A or B' and 'A and B' are not equivalent. Their truth tables are different.

Negation of 'and' and 'or'

The negations of 'and' and 'or' have a neat symmetry to them:

'not $(A$ and $B)$' is equivalent to '(not A) or (not B)',
'not $(A$ or $B)$' is equivalent to '(not A) and (not B)'.

To see the first, consider the negation of the statement

'Carl is tall and Carl is dark haired.'

We can see that at least one of the statements 'Carl is tall', 'Carl is dark haired' has to be false, so we can see at least one of 'Carl is not tall', 'Carl is not dark haired' is true. In other words, 'Carl is not tall or Carl is not dark-haired' is true. (Remember that this includes the case where both are true!)

We can apply similar reasoning to an example of the negation of 'or'. Create your own example to show this.

More generally, we can show that 'not $(A$ and $B)$' is equivalent to '(not A) or (not B)' using truth tables.

A	B	A and B	$\text{not}(A)$	$\text{not}(B)$	$(\text{not}(A))$ or $(\text{not}(B))$	$\text{not}(A$ and $B)$
F	F	F	T	T	T	T
F	T	F	T	F	T	T
T	F	F	F	T	T	T
T	T	T	F	F	F	F

The last two columns give 'not $(A$ and $B)$' and '(not A) or (not B)'. Since the columns are the same we can conclude that the statements are equivalent.

Exercise 6.10

Construct a truth table to show that 'not (A or B)' is equivalent to '(not A) and (not B)'.

Exercises

Exercises 6.11

(i)   Which of the following are statements? Are those that are statements true?
   (a) Life is sweet.
   (b) Is 2 a prime?
   (c) Prove that 2 is a prime.
   (d) The President of the United States in 1789 was a man.
   (e) The President of the United States in 2089 will be a woman.

(ii) Construct truth tables for the following:
   (a) not(A and B),
   (b) not(A or B),
   (c) (not A) or (not B),
   (d) A or (not B),
   (e) (not B) or B,
   (f) (not B) and B.

(iii) Negate the following:
   (a) A is true or B is false.
   (b) A is false and B is true.
   (c) A is true or B is true.
   (d) A is true and B is true.

(iv) It is simple to create truth tables with many inputs. Construct truth tables for the following:
   (a) A and (B or C),
   (b) (A and B) or C,
   (c) (not(A or B)) and C.

(v)   A **tautology** is a statement such that the truth table has True for all outputs. A **contradiction** is a statement such that the truth table has False for all outputs.
   (a) Show that 'A or not A' is a tautology.
   (b) Create a tautology using the 'and' connective.
   (c) Is the phrase 'Winners don't quit and quitters don't win' a tautology?
   (d) Show that 'A and not A' is a contradiction.
   Contradiction will be very useful in Part IV.

Summary

▶ A statement is a sentence that is either true or false – but not both.
▶ Law of excluded middle: statements are true or false; there is no in-between.

▶ The negation of statement A is the statement that is false when A is true.

▶ Statements can be constructed using the connectives 'or' and 'and'.

▶ Two statements are equivalent if they have the same truth table.

▶ 'not(A and B)' is equivalent to '(not A) or (not B)'.

▶ 'not(A or B)' is equivalent to '(not A) and (not B)'.

CHAPTER 7

Implications

Mathematics consists of propositions of the form: P implies Q,
but you never ask whether P is true.
Bertrand Russell

Russell's quote above is extremely incisive. Modern mathematics is indeed made up of statements of the form statement P implies statement Q. That is, we have 'If statement P is true, then statement Q is true also.' Usually, however, this structure is hidden, mainly to make mathematics more comprehensible – it would be hard to read if we always wrote it that way.

The second part of Russell's quote is also true but a lot more subtle. One could argue that the statement 'The Moon is made of cheese implies the Moon is a tasty snack' is true because if the Moon was cheese, then it would be tasty. The point is that the statement makes sense and is true yet it has nothing to say on whether the Moon really is made of cheese or whether it really is a tasty snack. All it says is that *if* it is cheesy, then it is tasty. It is worth bearing this example in mind as we proceed.

Instead of Russell's P and Q we will, in general, use A and B to denote our statements.

'If ..., then ...' statements

Most mathematical statements are of the form

'If statement A is true, then statement B is true.'

They may be heavily disguised but when you break them down, that is what you will find. This type of statement is called an **implication**. We say A implies B and sometimes write $A \Longrightarrow B$. Please refer to page 37 concerning the correct use of this symbol.

Examples 7.1

(i)   If I am Winston Churchill, then I am English.
(ii) If I am English, then I am Winston Churchill.
(iii) If I am President George Washington, then I am the first President of the United States of America.
(iv) If $a < b$, then $a^2 < b^2$.

(v)   If $a < c$, then $a < b$.

(vi)   If x is even, then x^2 is even.

Statements (i) and (iii) are true. Statement (ii) is not true – take an English person who is not Winston Churchill. The sentences (iv) and (v) are conditional upon a, b and c (so aren't really statements in the strict sense of the word). Statement (vi) is true as you probably know. We will prove it later.

Implication statements may be in a disguised form. For example, consider

'The sum of two even numbers is even.'

This can be rewritten as

'If x and y are even numbers, then $x + y$ is an even number.'

Similarly, the statement

'All prime numbers greater than two are odd.'

can be rewritten as

'If p is a prime number and $p > 2$, then p is odd.'

Hypothesis, assumption and conclusion

In an implication $A \Longrightarrow B$ there are two parts:

• Statement A is called the **hypothesis** or **assumption**, and
• Statement B is called the **conclusion**.

Thus, we are saying 'If the hypothesis is true, then the conclusion is true.'

Sometimes statement A will be made up of a number of sentences and so we refer to these as the hypotheses, assumptions and also call them conditions.

In the examples above 'I am George Washington' and '$a < b$' are examples of assumptions. Note that in (v) the statement '$a < b$' is also a conclusion.

Sometimes the assumptions and conclusions may be harder to spot.

'Suppose that x is a positive natural number. If x is odd, then x^2 is odd.'

What are the assumptions here? They are 'x is a positive natural number that is odd'. Another way of writing this is

'x is a positive odd number $\Longrightarrow x^2$ is odd.'

As you can see there are many ways of writing the same thing.

Exercises 7.2

(i) Identify the hypotheses and conclusions in the following.

(a) If $x^2 > 5$, then $x^3 + 12x + 4 < 3$.

(b) We have $x^2 + 1 > 3$ if $x < \sqrt{2}$.

   (c) Let T be a right-angled triangle. If c is the hypotenuse, then $c^2 = a^2 + b^2$, where a and b are the lengths of the other two sides. (Careful here! The assumption is *not* 'c is the hypotenuse'.)

   (d) Let T be a triangle. Suppose that T has edges of length a, b and c with the angle opposite a equal to θ. Then $a^2 = b^2 + c^2 - 2bc\cos\theta$.

The truth of A and B when A implies B

Let us return to the idea behind Russell's quote at the start of the chapter and consider the example from the introduction: 'The Moon is made of cheese implies the Moon is a tasty snack.' The assumption is that the Moon is made of cheese and the conclusion is that it is a tasty snack. Hopefully we can agree that neither of these statements is true, but can agree that the *implication* is true.

This leads us to the most fundamental fact to grasp about implication:

> '$A \Longrightarrow B$' says *nothing* about the truth or otherwise of A or B.

I cannot stress enough how it is vital to grasp this.

Supposing that $A \Longrightarrow B$ is true we have three possibilities:

(i) A is true and B is true.

(ii) A is false and B is false.

(iii) A is false and B is true.

We cannot have A is true and B is false, because $A \Longrightarrow B$ *means* that if A is true, then B is true.

The first possibility above can occur: If I am Winston Churchill, then I am English.

The second possibility can also occur. Let A be $1 = 2$ and B be $5 = 6$. Then we can see $A \Longrightarrow B$ is true because if $1 = 2$ is true and we add 4 to both sides we see that $1 + 4 = 2 + 4$, i.e. $5 = 6$. So this possibility can exist.

The third possibility seems to be counterintuitive. Many beginners assume that false statements can never imply true statements, but this is not the case and as it is such a common fallacy it is the subject of a separate section.

False statements can imply true statements

We have already noted that '$A \Longrightarrow B$' says nothing about the truth of A or B. However, it is worth explicitly noting that false statements can imply true statements. For example, suppose that $-1 = 1$; obviously it doesn't, but we are mathematicians and we can suppose what we like. Squaring both sides, we get $1 = 1$, which *is* true. Thus,

$$'-1 = 1 \Longrightarrow 1 = 1'$$

is *true*! It does not matter that $-1 = 1$ is false.

In fact, using just one false statement we can prove anything! We will leave that to books on logic.

Truth table for $A \Longrightarrow B$

What should the truth table of implication look like? It is the following:

A	B	$A \Longrightarrow B$
F	F	T
F	T	T
T	F	F
T	T	T

The final two lines should be obvious. (Think like a mathematician! Check the text!) The other two are harder to understand. In some sense they are a *definition* of the truth table. On the other hand, they can make sense because if A is false, then B is allowed to be anything, so we might as well say $A \Longrightarrow B$ when A is false.

Exercise 7.3

Show that the truth tables for $A \Longrightarrow B$ and 'not(A) or B' are the same.

Negation of 'If …, then …'

Since $A \Longrightarrow B$ is a statement we can ask what its negation is. The obvious response is to say that A does not imply B. What does this mean? And what is the correct logical statement? It means that A is true and at the same time B is false. This is 'A and not(B)'.

We can check this using Exercise 7.3. There it is shown that '$A \Longrightarrow B$' is the same as 'not(A) or B'. Recall from Chapter 6 that the negation of 'P and Q' is 'not(P) or not(Q).' Thus, the negation of '$A \Longrightarrow B$' is the negation of 'not(A) or B.' This is 'not(not(A)) and not(B)', which simplifies to 'A and not(B).'

Exercise 7.4

Express the preceding argument using truth tables.

Remark 7.5

Note that the negation of an implication is not another implication.

Different ways of writing A implies B

'B if A' is the same as '$A \Longrightarrow B$'

It is obvious that 'If I am Winston Churchill, then I am English' can be rewritten as 'I am English if I am Winston Churchill.' We can do this more generally. 'If A, then B' can be rewritten as

'B if A'.

Occasionally, students get confused because they are used to thinking of B being on the right (or at the end) as in '$A \Longrightarrow B$' and 'If A, then B.' Thus if you are used to thinking of the assumptions on the left and the conclusion on the right you will be in trouble. This way of writing will become crucial when we look at equivalence of statements in Chapter 9.

'*A* only if *B*' is the same as '*A* $\Longrightarrow$ *B*'

The next way of writing '*A* $\Longrightarrow$ *B*' is less obvious. It replaces 'If I am Winston Churchill, then I am English' with 'I am Winston Churchill only if I am English.' The idea is that if I am not English, then there is no way I could be Winston Churchill.

So, more generally, another way of writing '*A* $\Longrightarrow$ *B*' is

  '*A* only if *B*'.

This is harder to accept as being equivalent to '*A* $\Longrightarrow$ *B*' so let's go through it slowly.

The statement '*A* only if *B*' says that *A* is true only if *B* is true. How is this the same as *A* is true implies *B* is true?

Suppose that '*A* $\Longrightarrow$ *B*' is true. Now suppose further that *A* was true and *B* was false, then this would mean that '*A* $\Longrightarrow$ *B*' was not true. Hence *A* can be true only if *B* is true. Similarly, suppose that *A* is true only if *B* is true. Then if *A* is true, we must have *B* is true, so *A* $\Longrightarrow$ *B*.

Exercise 7.6

Explain the above using truth tables, in particular, show using a truth table that '*A* $\Longrightarrow$ *B*' is the same as '*A* only if *B*'.

Exercises

Exercises 7.7

(i)   Suppose that *A* is true and *B* is false. Which of the following are true?
   (a) $A \Longrightarrow B$,   (b) $B \Longrightarrow A$,   (c) not $B \Longrightarrow A$,
   (d) $A \Longrightarrow A$,   (e) *A* or not *B*,   (f) not $A \Longrightarrow A$,
   (g) not(*A* or *B*).

(ii)   Negate the following:
   (a) If you score 70%, then you have done well in this course.
   (b) If it rains, then I will stay at home.
   (c) If $x^2 + 2x + 1 = 0$, then $x = -1$.
   (d) $x^2 + x - 2 = 0$ implies $x = 1$ or $x = -2$.
   (Hint. The negation of an 'if…, then…' is not another 'if…, then…')

(iii)   Write the following as statements using 'only if':
   (a) If $x = -2$, then $x^2 = 4$.
   (b) If *x* and *y* are odd, then xy is odd.
   (c) $x^2 + x - 2 = 0 \Longrightarrow x = 1$ or $x = -2$.
   (d) $x^2 + x - 2 = 0 \Longleftarrow x = 1$.

(iv)   Construct truth tables for
   (a) $(A \Longrightarrow B) \Longrightarrow$ not *A*, and

(b) A or $(B \Longrightarrow$ not $A)$.

(v) Construct truth tables to show that

(a) $A \Longrightarrow (A$ or $B)$ and

(b) $[(A$ and $B) \Longrightarrow C] \Longrightarrow [$not $C \Longrightarrow ($not A or not $B)]$

are tautologies.

Summary

▶ In the statement '$A \Longrightarrow B$' statement A is called the hypothesis or assumption and statement B is called the conclusion.

▶ '$A \Longrightarrow B$' can be written as 'If A, then B.'

▶ 'A implies B' means that if A is true, then B is true. Nothing else. If A is false, then B may be true, or it may be false.

▶ The negation of $A \Longrightarrow B$ is the same as 'not(A) or B'.

▶ 'A only if B' is the same as '$A \Longrightarrow B$'.

▶ 'B if A' is the same as '$A \Longrightarrow B$'.

CHAPTER 8

Finer points concerning implications

He's suffering from Politician's Logic. Something must be done, this is something, therefore we must do it.
Antony Jay and Jonathan Lynn,
Power to the People, episode of *Yes, Prime Minister*

The English language is full of ambiguities. The humorous quote above exploits this and works because the word 'something' has two different meanings in the sentence. With mathematical language we aim to remove such ambiguities. In this chapter we will first look at a particular problem that can occur in everyday usage with the use of the if/then structure. We shall see how this example leads us to consider what are called the inverse statement and the contrapositive statement. This shows that although using everyday examples can be very illuminating we do have to be careful as the English language can play tricks on us.

The inverse: a common mistake

The most common initial mistake that people make in logic arises from an everyday usage we learn at a young age. For example, an adult says to a child

'If you don't tidy your room, then you won't get ice-cream.'

Our instinct, like the child and parent, is to interpret this as a contract: if the child tidies their room, then they get ice-cream. In other words

'If you tidy your room, then you will get ice-cream.'

Yet the original statement does not say that. It only says what will happen if the child does *not* tidy their room. It says nothing about what happens if they do! Yet I am sure most of us would agree that if the child tidied their room, and did *not* get ice-cream, they would have reason to feel aggrieved.

Another example is 'If it rains, then I won't go.' We know in everyday speech that this implicitly means that if *doesn't* rain, then I *will* go. However, strictly speaking, there is only a statement about what will happen if it rains; it does not tell us anything about what I will do if there is no rain.

The point is that we grow up with certain ideas about what if/then sentences mean. Let us make the problem with the above examples explicit by considering the following example.

'If I am Winston Churchill, then I am English.'

This statement is true as we established earlier that Winston Churchill was English. Now let's add some negations so that the statement becomes

'If I am not Winston Churchill, then I am not English.'

This is definitely not true, since it is possible to find someone who is not Winston Churchill yet is English.

Abstracting this to mathematical language we make the following definition.

Definition 8.1

*The **inverse** of the implication 'If A, then B' is*

'If not(A), then not(B).'

Example 8.2

In the room/ice-cream example above we let A be 'you don't tidy your room' and B be 'you won't get ice cream', and so 'not(A)' is 'you tidy your room' and 'not(B)' is 'you will get ice-cream'. The inverse statement of 'If you don't tidy your room, then you won't get ice-cream' is thus 'If you tidy your room, then you will get ice-cream.'

The Churchill/English example above shows that if 'If A, then B' is true, then 'If not(A), then not(B)' can be false, so we can see, logically speaking, the truth of the initial statement about rooms and ice-cream and its inverse can be totally different. (However, the child's experience will be that they are equivalent.)

The distinction we need is:

'$A \implies B$' is not equivalent to 'not(A) $\implies$ not(B)'.

The Churchill/English example is good for remembering this distinction as it shows that an implication can be true while its inverse is false.

Note that an implication and its inverse *can* both be true as the following example shows.

Example 8.3

Consider the statement

'x is even implies that x^2 is even'.

This is true. You may like to show that this is the case!

The inverse statement is

'x is not even implies that x^2 is not even'.

In other words

'x is odd implies that x^2 is odd'.

This is also true! Thus, in this case 'If A, then B' and 'If not(A), then not(B)' are both true.

Considering this example and the Churchill example we see that:

> In general, the truth of '$A \Longrightarrow B$' says nothing about the truth of its inverse, 'not(A) $\Longrightarrow$ not(B)'.

Dangerous examples

The simple room/ice-cream example is fairly innocuous. Little harm is done by it. Let's see how this sloppy logic can be used in real life and get us into potentially explosive situations with other people. Consider the statement

'If you are a mathematician, then you are intelligent',

and assume that it is true.

Note that this says nothing about the intelligence of non-mathematicians. A person could be a geographer, historian or telephone engineer but this statement won't tell you anything about them. Yet people may be offended as they will believe that it insults non-mathematicians: 'Hey, I'm a telephone engineer! Are you saying I'm stupid?'

To see how this could become very controversial, imagine that rather than using 'mathematician' I had said 'If you are a man, then you are intelligent' and asserted its truth. Many would argue against this, not by pointing out that the logic is wrong – there are plenty of unintelligent men around – but by complaining that it was sexist and arguing that there are plenty of intelligent women around. Of course, rarely do people make such statements in this form; usually it is dressed up in some manner.

Necessary and sufficient

Logical concepts are also often stated in terms of necessary and sufficient conditions. Fortunately, the terms 'necessary' and 'sufficient' follow the standard everyday usage.

Necessary conditions

A necessary condition is one that is needed for something to be true.

Definition 8.4

*A **necessary** condition is one which must hold for a conclusion to be true. It does not guarantee that the result is true.*

In other words '*A* is necessary for *B*' means that '*B* is true only if *A* is true'. We know that this latter is the same as $B \Longrightarrow A$.

Examples 8.5

(i)   Being an article of clothing is necessary for being a glove.
(ii) $x \leq 0$ is necessary for $x = -2$.
(iii) For a triangle to be an equilateral triangle (all angles equal) it is necessary that it be an isosceles triangle (two angles equal).

Sufficient conditions

A sufficient condition is one that is enough to guarantee a statement is true.

Definition 8.6

*A **sufficient** condition is one which guarantees the conclusion is true. The conclusion may be true even if the condition is not satisfied*

In other words '*A* is sufficient for *B*' means '*B* is true if *A* is true'. This is the same as $A \Longrightarrow B$.

Examples 8.7

(i)   Being a glove is sufficient for being an article of clothing. Note that being a glove is not necessary for being an article of clothing.
(ii) $x = -2$ is sufficient for $x \leq 0$.
(iii) For a triangle to be an isosceles triangle it is sufficient that it be an equilateral triangle. Note that the triangle does not have to be isosceles to be equilateral.

The contrapositive

We have seen that $A \Longrightarrow B$ is not equivalent to $\text{not}(A) \Longrightarrow \text{not}(B)$ even though everyday language has trained us to consider it true. Worse still, language does not prepare us for a result that is true.

Definition 8.8

*The **contrapositive** of the statement '$A \Longrightarrow B$' is*

> '$\text{not}(B) \Longrightarrow \text{not}(A)$'.

Surprisingly this is equivalent to $A \Longrightarrow B$! That is, we can replace $A \Longrightarrow B$ by $\text{not}(B) \Longrightarrow \text{not}(A)$.

Let's see some examples.

Examples 8.9

(i)   'If I am Winston Churchill, then I am English' has contrapositive 'If am not English, then I am not Winston Churchill.'

(ii) 'I am not American implies that I am not a Texan' has contrapositive 'If I am a Texan, then I am American.'

(iii) 'I am Jane implies that I am a woman' has contrapositive 'If I am not a woman, then I am not Jane.'

(iv) '$x^2 - 4x - 5 = 0 \Longrightarrow x \geq -2$' has contrapositive '$x < -2 \Longrightarrow x^2 - 4x - 5 \neq 0$.'

Exercises 8.10

Find the contrapositive statements for the following:

(i)   If $x = 2$, then $\sqrt{x}$ is irrational.

(ii) If x is prime, then $x = 2$ or x is odd.

(iii) If S is a square, then S is a rectangle.

(iv) A circle is an ellipse.

Exercise 8.11

Construct a truth table for $\text{not}(B) \Longrightarrow \text{not}(A)$ and show it is the same as the one for $A \Longrightarrow B$.

Remark 8.12

The previous exercise shows that a statement and its contrapositive are equivalent.

Exercises

Exercises 8.13

(i)   Rewrite the following as 'if…, then …' statements.

  (a) A sufficient condition for Peter to win the Championship is that he wins in Brazil.

  (b) A necessary condition for Stuart to win the Championship is that he beats Peter.

  (c) Regular work is sufficient to pass the course.

  (d) Regular work is not necessary to pass the course.

  (e) To be President of the United States of America it is necessary to be born in the USA.

  (f) To be Prime Minister of India it is not necessary to be born in India.

(ii) Let A be '$x^2 - 2x - 3 > 0$' and B be '$x > 3$'. Which of the following are true and which are false?

  (a) $A \Longrightarrow B$,   (b) $B \Longrightarrow A$,

  (c) A is necessary for B,   (d) B is sufficient for A,

  (e) A is sufficient for B,   (f) B is necessary for A,

  (g) $\text{not}(B) \Longrightarrow \text{not}(A)$,   (h) $\text{not}(A) \Longrightarrow \text{not}(B)$.

(iii) Which of the following statements are equivalent?

  (a) If my team lost the last game, then they must have lost the championship.

  (b) If my team lost the last game, then your team won the championship.

(c) If my team lost the last game, then they won the championship.

(d) If my team won the championship, then they won the last game.

(e) If my team won the last game, then they won the championship.

(f) If my team lost the championship, then they must have lost the last game.

(iv) What is the contrapositive of $A \Longrightarrow (B \Longrightarrow C)$?

(v) We now come to an interesting experiment conducted by P.C. Wason on understanding. (A lot of people get this wrong.)

Suppose that students were told that on each side of a card was a vowel and on the other was a number. The students might be shown the cards:

The task was to decide whether the rule 'If a card has a vowel on one side, then it has an even number on the other side' was true or not. The students had to turn over only the cards which had to be turned over to judge the correctness of the rule.

Which cards in the above example would you turn over and which would you not? Justify you answers.

Summary

▶ The inverse of '$A \Longrightarrow B$' is 'not$(A) \Longrightarrow$ not(B)'.

▶ '$A \Longrightarrow B$' true **does not** mean that 'not$(A) \Longrightarrow$ not(B)' is true.

▶ 'A is necessary for B' is equivalent to '$B \Longrightarrow A$'.

▶ 'A is sufficient for B' is equivalent to '$A \Longrightarrow B$'.

▶ The contrapositive of '$A \Longrightarrow B$' is 'not$(B) \Longrightarrow$ not(A)'.

▶ An implication and its contrapositive are equivalent.

CHAPTER 9

Converse and equivalence

But the fact that some geniuses were laughed at does not imply that all who are laughed at are geniuses. They laughed at Columbus, they laughed at Fulton [steamboat inventor], they laughed at the Wright brothers. But they also laughed at Bozo the Clown.
Carl Sagan, *Broca's Brain*, 1979

Statements of the form $A \implies B$ are at the heart of mathematics. We have seen that for an implication $A \implies B$ we can take its inverse $(\text{not}(A) \implies \text{not}(B))$ and its contrapositive $(\text{not}(B) \implies \text{not}(A))$. In this chapter we will look at another implication: $B \implies A$; this is called the converse of $A \implies B$. We shall see that a statement and its converse are not the same. One may be true and the other false, both may be true or both may be false.

If $A \implies B$ and $B \implies A$ are both true, then we say that A and B are equivalent statements. Mathematicians really like equivalent statements, particularly if the A and B seem to have no obvious connection.

The converse

Definition 9.1
*The **converse** of the statement '$A \implies B$' is '$B \implies A$'.*

The converse of

'If I am Winston Churchill, then I am English'

is

'If I am English, then I am Winston Churchill.'

This simple example shows that, even if a particular statement is true, its converse need not true. It may be true or it may not be true. Investigation is required before we can say.

Exercise 9.2

Find the converse of each of the following:

(a) If my cat is black, then not all cats are grey.
(b) If x is even, then x^2 is even.
(c) If I am a Corsican, then I am Napoleon.

A good mathematician, when presented with an A implies B type statement, will ask 'Is the converse true?' Internalize this question and make it part of your tool kit for doing mathematics. Whether the converse is true or not is not too important, the point is that the exercise sharpens mathematical ability.

In one of my lecture courses for first-year students I offer a prize for the first student to ask this question during a lecture. To win the prize they have to ask it at a point where the converse is interesting and not obviously true or obviously false. I also allow them to ask it during other the courses of my colleagues since I want my students to use their knowledge outside my classes.

Exercise 9.3

What is another name for the inverse of the converse?

Logical equivalence

We have talked about two statements being the same, that is, equivalent. For example, $\text{not}(\text{not}(A))$ is the same as A, or $\text{not}(B) \implies \text{not}(A)$ is the same as $A \implies B$, and have used truth tables to show this. We will now investigate this notion of equivalence in more detail.

Example 9.4

Consider the statements 'I am President George Washington' and 'I am the first president of the United States of America.'

We know that 'If I am President George Washington, then I am the first president of the United States of America' is true. Its converse 'If I am the first president of the United States of America, then I am President George Washington' is also true.

This is because being George Washington and being the first president of the United States are the same thing; they are equivalent. Let's put this concept into mathematical terminology.

Definition 9.5

*If $A \implies B$ and $B \implies A$ are both true, we say that A and B are **logically equivalent statements** and write $A \iff B$. The sign $\iff$ is read as 'if and only if' and the statement is called a **biconditional statement**.*

The reason for using 'if and only if' is that we know that '$A \implies B$' can be written as 'A only if B' and '$B \implies A$' can be written as 'A if B'. Combining the two implications we get the 'if and only if'. However, students often have much confusion about which implication is the 'if' and which implication the 'only if'. Practise spotting them.

Sometimes people use 'iff' as an abbreviation of 'if and only if'. (But to me it always looks like a typographical error!)

Example 9.6

Consider the following statement:

$$x^2 + 2x - 3 = 0 \Longleftrightarrow x = 1 \text{ or } x = -3.$$

The 'only if' (i.e. $\Rightarrow$) part of the statement is $x^2 + 2x - 3 = 0 \Longrightarrow x = 1$ or $x = -3$. This is true. To see this solve the equation via your preferred method. Here, it is easy to factorize the quadratic: $x^2 + 2x - 3 = (x - 1)(x + 3)$, so $x^2 + 2x - 3 = 0$ implies $x = 1$ or $x = -3$.

The 'if' (i.e. $\Leftarrow$) part of the statement is $x = 1$ or $x = -3 \Longrightarrow x^2 + 2x - 3 = 0$. This is true as well. This is easy to check:

$$1^2 + 2 \times 1 - 3 = 1 + 2 - 3 = 0,$$
$$(-3)^2 + 2(-3) - 3 = 9 - 6 - 3 = 0.$$

Thus the two statements, '$x^2 + 2x - 3 = 0$' and '$x = 1$ or $x = -3$', are equivalent.

Example 9.7

Now consider what happens for the statement

$$x^2 + 2x - 3 = 0 \Longleftrightarrow x = -3.$$

From the previous example we know that this statement is false. (Because '$x^2 + 2x - 3 = 0$' is equivalent to '$x = 1$ or $x = -3$' and this latter statement is obviously not equivalent to $x = -3$.)

Ok, it's false but which 'direction' goes wrong? Is it the $\Leftarrow$ or the $\Rightarrow$ part that is false? That is, is it the 'if' or 'only if'?

It must be the $\Rightarrow$ direction, i.e. the 'only if', as $x^2 + 2x - 3 = 0$ has more solutions that just $x = -3$.

An aside

When presented with a question concerning solutions of equations, say, show that $x = 4$ is a solution of $3x^2 - 14x + 8 = 0$, many students will *solve* the equation by some method, hence producing $x = 4$ and $x = 2/3$ as solutions. Yet it is much simpler than that. To show something is a solution we just put it back into the equation: $3x^2 - 14x + 8 = 3(4)^2 - 14(4) + 8 = 48 - 56 + 8 = 0$.

Another example

In the next example we shall see an 'if and only if' statement and show that it is true.

Example 9.8

The integer n is even if and only if $n + 1$ is odd.

We can show that this statement is true. First we show that n is even only if $n + 1$ is odd. Assume n is even. Then $n = 2m$ for some $m \in \mathbb{Z}$. Thus $n + 1 = 2m + 1$, which is odd by definition. Therefore $n + 1$ is odd.

Now we show that n is even if $n + 1$ is odd. Assume that $n + 1$ is odd, i.e. $n + 1 = 2m + 1$ for some $m \in \mathbb{Z}$. Then $n = 2m$ by cancellation. Hence, n is even.

The point of this example is to show that there are two statements to be shown true. One statement involves the 'if' and the other involves the 'only if'. Much more will be said about this later in the book.

Exercises

Exercises 9.9

(i)   Write the converse of each of the following statements:
    (a) If $x > 5$, then X is red.
    (b) An integer can be even or odd but it cannot be both.
    (c) Eating ice-cream is necessary for me to be happy all day.
    (d) Eating ice-cream is sufficient for me to be happy all day.
    (e) It is not necessary to understand things to argue about them.
    (f) Stop, or I will shoot!

(ii) Which of the following four statements are equivalent?
    (a) If A and B are both red, then X is true.
    (b) If A and B are both not red, then X is true.
    (c) If X is false, then A and B are not both red.
    (d) If A or B is not red, then X is true.
    If A is red and B is yellow, then in (a) is X true? Is it false? What about in (d)?

(iii) Create a truth table for
    (a) $A \Longleftrightarrow B$, and for
    (b) the converse of 'A implies B'.

(iv) Give the converse of each of the following statements. Combine the statement and its converse into a biconditional statement. (Note the statements do not have to be true!)
    (a) If $ab = 0$, then $a = 0$ or $b = 0$.
    (b) If $2x^2 - 7x + 6 = 0$, then $x = 2$.
    (c) The product of two odd integers is odd.
    (d) The sum of an odd number and an even number is odd.
    (e) Suppose that A, B and C are sets. In this case, $A \cap (B \cup C) \subseteq (A \cap B) \cup (A \cap C)$.
    (f) Suppose that f is polynomial in the variable x. If $f(a) = 0$, then $x - a$ is a factor of f.

Summary

▶ Using statements A and B, we can combine not and $\implies$ in four ways:

  (i) $A \implies B$,

  (ii) $B \implies A$, the converse,

  (iii) not $A \implies$ not B, the inverse,

  (iv) not $B \implies$ not A, the contrapositive.

▶ If $A \implies B$ is true, then the contrapositive is true.

▶ If $A \implies B$ is true, then the inverse and converse may be true, or they may be false.

▶ If $A \implies B$ and $B \implies A$, then we say that A and B are equivalent and write $A \iff B$.

▶ 'iff' is an abbreviation of 'if and only if'.

CHAPTER 10

Quantifiers – For all and There exists

You can fool all the people some of the time, and some of the people all the time,
but you cannot fool all the people all the time.
Abraham Lincoln

In Chapter 6, Making a statement, we called a sentence like 'x is an odd number' a conditional statement. Whether or not it was true depended on x. In this chapter we shall describe two fundamental ways of quantifying x so that we get statements rather than conditional statements.

We can exemplify them via the following. The sentence '$x^2 = 2$' is conditional on x. By combining such sentences with some description of the x we can form statements. For example, 'For all $x \in \mathbb{Z}$, $x^2 = 2$' or 'There exists an $x \in \mathbb{R}$ such that $x^2 = 2$.' Note that the latter is true and the former is false.

Since we are giving a description of how many of x we are talking about, i.e. assigning a quantity, we call these descriptions **quantifiers**. The fancy titles for the quantifiers we are interested in are universal quantifier and existential quantifier.

For all – the universal quantifier

Definition 10.1

*The phrase 'for all' is the **universal quantifier**. It is denoted $\forall$. (This is an upside down A. Think of the A in 'All' to remember this.)*

Let's see this in action.

Examples 10.2

(i)   'For all $x \in \mathbb{R}$, $x^2 \geq 0$.' This says that the square of a real number is positive.

(ii)   Let S be a subset of $\mathbb{R}$ and U be a real number. 'For all $s \in S \subseteq \mathbb{R}$, $s \leq U$.' This says that every element of a set S is less than or equal to the number U.

(iii) Let O be the set of odd numbers. 'For all $x \in O$, $x^2 + 1$ is odd.' This says that, for odd x, the number $x^2 + 1$ is also odd. (This is false, by the way.)

(iv)   'For all rational numbers x and y, the product xy and sum $x + y$ are rational.'

For a sentence $P(x)$, such as $x^2 \geq 0$, we can write a statement in the form $\forall x \in \mathbb{R}$, $P(x)$, e.g. $\forall x \in \mathbb{R}$, $(x^2 \geq 0)$. Note that here we enclosed $x^2 \geq 0$ in brackets to make it clear that this is the sentence being quantified. If the sentence has two variables being quantified, then we can use the notation $P(x, y)$ for the sentence.

We can rewrite example (iv) in symbols as:

$$\forall x \in \mathbb{Q} \, \forall y \in \mathbb{Q}(xy \in \mathbb{Q} \text{ and } x + y \in \mathbb{Q})$$

or as

$$\forall x, y \in \mathbb{Q}(xy \in \mathbb{Q} \text{ and } x + y \in \mathbb{Q}).$$

We can have statements with implications: 'For all real numbers x, if $x \geq 3$, then $x^2 \geq 9$.' We can rewrite this as '$\forall x \in \mathbb{R}$, $(x \geq 3 \implies x^2 \geq 9)$.'

Handy tip 10.3

It is a good idea to translate statements written in words to statements written in symbols and vice versa.

There are other ways of writing 'for all', we can use 'every', 'for every', 'for each'. For example, 'For every $x \in \mathbb{R}$, $x \geq 0$.'

One synonym of 'for all' in the English language is 'for any'. Unfortunately its use in English is ambiguous and so mathematicians avoid it. Consider the following exercise: 'Show that $x^2 \geq 0$ for any $x \in \mathbb{Z}$.' This can be interpreted as an instruction to show 'For all $x \in \mathbb{Z}$, $x^2 \geq 0$.' However, it can also be interpreted as saying 'Choose a single integer and show that $x^2 \geq 0$.' This is easy to do! Take $x = 3$, then $x^2 = 9 \geq 0$. These two interpretations are obviously not the same yet are reasonable responses to the use of the word 'any'.

There exists – the existential quantifier

Definition 10.4

*The phrase 'there exists' is the **existential quantifier**. It is denoted $\exists$. (This is a backward E. Think of the E in 'Exists' to remember it.)*

Examples 10.5

(i)   'There exists $x \in \mathbb{Z}$ such that $x^2 = 4$.' This is true as $x = 2$ satisfies $x^2 = 4$. Note that $x = -2$ also satisfies the equation. Thus, saying there exists an x does not mean that there is only one such x.

(ii) 'There exists $x \in \mathbb{Z}$ such that $x^2 = 5$.' This is not true.

(iii) '$\exists x \in \mathbb{Z}(x^2 - 4x + 3 = 0)$.' Again true, just solve the equation to see this.

When we read examples like the last one, we insert a 'such that' between the quantifier and the sentence to which it refers. So we read 'There exists an integer x such that $x^2 - 4x + 3 = 0$.'

Combining the quantifiers

It is possible to combine the two quantifiers.

Example 10.6

'For all $x \in \mathbb{Z}$ there exists $y \in \mathbb{Z}$ such that $y > x$.' This can be written in symbols as '$\forall x \in \mathbb{Z} \exists y \in \mathbb{Z}(y > x)$.' This just says that for each integer we can find a bigger one. Given any integer x, I can find a y bigger than it. For example if you give me $x = 15$, then I can take $y = 16$, or 20, or 5 000 003.

We can also combine the quantifiers in definitions.

Example 10.7

Let S be a subset of the real numbers $\mathbb{R}$. We say that S has an **upper bound** if there exists U such that for all $s \in S, s \leq U$. We can write this as 'We say that S has an **upper bound** if $\exists U \, \forall s \in S(s \leq U)$.'

This just says that every element of a set S is less than or equal to the number U. For example, if set $S = \{-10, 4, 13, 17\}$, then $U = 20$ is an upper bound and so is $U = 17$, but $U = 6$ is not. The set $\mathbb{N}$ has no upper bound.

Remark 10.8

To help with reading statements we place the quantifier before the statement to which it refers. Thus '$\forall x \in \mathbb{R} \exists y \in \mathbb{Z}(y > x)$' is really '$\forall x \in \mathbb{R}, P(x)$' where $P(x)$ is the sentence '$\exists y \in \mathbb{Z}(y > x)$'.

Writing the quantifier first is not standard in English usage, only when using the symbolic notation $\forall$ and $\exists$. Furthermore, many mathematicians, when writing in words put 'for all' at the end (or, worse, miss it out altogether). For example, '$x^2 \geq 0$ for all $x \in \mathbb{R}$.'

Warning! The order of quantifiers is important

The order of the quantifiers is vital to the meaning of a statement. Consider the earlier example, 'For all $x \in \mathbb{Z}$ there exists $y \in \mathbb{Z}$ such that $y > x$', which we decided was true since for any integer x you give me I can find a bigger integer. The statement can be written symbolically as $\forall x \in \mathbb{Z} \exists y \in \mathbb{Z}(y > x)$.

Now, reversing the quantifiers we have $\exists y \in \mathbb{Z} \forall x \in \mathbb{Z}(y > x)$. We read this as the statement 'There exists an integer y such that for all integers x we have $y > x$.' This is false. It says I can find an integer y so that if you give me an integer x, then my y is bigger than your x. This can't be true since if you came along with $y + 1$, then it would be an integer bigger than my y.

If you did not understand this subtlety, then I suggest you read again until you do – it will save you a lot of problems later!

Example 10.9

Recall the definition of upper bound above. The condition was 'We say that S has an **upper bound** if $\exists U \, \forall s \in S(s \leq U)$.' If we reverse the $\exists$ and $\forall$, then the condition becomes $\forall s \in S \, \exists U (s \leq U)$. This condition is always satisfied. Whatever s is chosen take $U = s + 1$, then $s \leq U$.

An important and helpful point is that in $\forall x \exists y P(x, y)$ the y can depend on the x.

Remark 10.10

Interchanging $\forall$ and $\forall$ or interchanging $\exists$ and $\exists$ does not change the meaning of a statement. Thus we may use $\forall x, y$ in place of $\forall x \, \forall y$, and $\exists x, y$ in place of $\exists x \, \exists y$.

Exercises

Exercises 10.11

(i) Rewrite the following using $\forall$ and $\exists$.
  (a) For all integers x, x is odd or even.
  (b) There exist two prime numbers such that their sum is prime.
  (c) There exists a rational number greater than $\sqrt{2}$.
  (d) If x is a real number, then x^2 is greater than x.
  (e) For all $n \in \mathbb{N}$ there exists a prime p such that $p > n$.

(ii) Decide whether the following are true or false. Explain your answers.
  (a) $\forall x \exists y (x^2 = y)$, where both x and y are in $\mathbb{R}$.
  (b) $\forall y \exists x (x^2 = y)$, where both x and y are in $\mathbb{R}$.
  (c) $\forall x \exists y (x^2 = y)$, where both x and y are integers.
  (d) $\forall y \exists x (x^2 = y)$, where both x and y are integers.
  (e) $\forall x \in \mathbb{R} \exists y \in \mathbb{R}(x + y = 0)$.
  (f) $\exists x \in \mathbb{R} \forall y \in \mathbb{R}(x + y = 1)$.
  (g) $\forall x P(x) \implies \exists x P(x)$.
  (h) $\exists x P(x) \implies \forall x P(x)$.
  (i) $\exists n \in \mathbb{N}$ such that $n^2 \leq n$.

Summary

► 'For all' is the universal quantifier, denoted $\forall$.
► 'There exists' is the existential quantifier, denoted $\exists$.
► Quantifiers can be combined but order is important.

CHAPTER 11

Complexity and negation of quantifiers

Everything is simpler than you think and at the same time more complex than you imagine.

Johann Wolfgang von Goethe

The more quantifiers a statement has the more complicated it is and the harder it is to understand. In this chapter we will look at how complicated statements can be made with quantifiers and give a method for seeing through this complexity.

We also see how to negate statements involving quantifiers. Fortunately, even for the most complicated of statements this is actually quite easy.

Complexity of quantifiers

The number of quantifiers in a mathematical statement gives a rough measure of the statement's complexity. Statements involving three or more quantifiers can be difficult to understand. This is the main reason why it is hard to understand the rigorous definitions of limit, convergence, continuity and differentiability in analysis as they have many quantifiers.

In fact, it is the alternation of the $\forall$ and $\exists$ that causes the complexity. For example, $\forall x \forall y \exists z P(x, y, z)$ will, in general, be simpler than $\forall x \exists z \forall y P(x, y, z)$. However, since we can replace $\forall x \forall y$ by $\forall x, y$, just counting the number of quantifiers gives a good measure of complexity.

Let's see some examples.

One quantifier

Examples 11.1

(i) $\forall x \in \mathbb{R}, x^2 \geq 0$.

(ii) The number U is an upper bound for $S \subseteq \mathbb{R}$ if, for all $s \in S \subseteq \mathbb{R}, s \leq U$.

(iii) There exists a solution to the equation $5x - \cos(4x) = 0$.

Two quantifiers

Examples 11.2

(i) $\forall x \exists y$ such that $y > x$.

(ii) $\exists y$ such that $\forall x \, (y > x)$.

(iii) 'There exists an upper bound for $S \subseteq \mathbb{R}$.' Note that here we have hidden a quantifier arising from the definition of upper bound. The sentence says

$$\exists U \in \mathbb{R} \text{ such that } \forall s \in S, \ s \leq U.$$

(iv) 'There exists a unique x such that $P(x)$ is true.' This means that there exists an x that satisfies $P(x)$, but more than that, if y is such that $P(y)$ is true, then y is equal to x. We can write this as

$$\exists x \, (P(x) \text{ and } \forall y (P(y) \implies y = x)).$$

Three and four quantifiers

Example 11.3

Suppose $a_1, a_2, a_3, \ldots$ is an infinite sequence of numbers. We call l the **limit of the sequence** if $\forall \epsilon > 0, \ \exists N \geq 1$ such that $\forall n \geq N, \ -\epsilon < a_n - l < \epsilon$.

That's quite a fierce definition. I'm not particularly interested in using it in this book; I am merely showing that three quantifier statements are possible. Now, given a sequence we can ask if l exists for that sequence and produce a statement involving four quantifiers.

Example 11.4

Suppose $a_1, a_2, a_3, \ldots$ is a sequence of numbers. The statement

$$\exists l \in \mathbb{R}, \ \forall \epsilon > 0 \, \exists N \geq 1 \text{ such that } \forall n \geq N, \ -\epsilon < a_n - l < \epsilon$$

says that the sequence has a limit.

Don't worry about understanding what these last two examples mean; just observe that even statements with three quantifiers are hard for the human brain to grasp. We need some way to see through this complexity. This is covered next.

The secret of seeing through the complexity

Here is a method that works for me when thinking about statements with many quantifiers.

I like to think of the quantifiers as being two different *processes*. In one someone hands me a random x and in the other I have to locate a particular x.

Statements beginning 'for all'

Statements involving 'for all' are of the form $\forall x, \, P(x)$ where $P(x)$ is some sentence, usually a condition. Here I imagine that someone *gives* me an x and I have to show that

$P(x)$ is true. But the person could give me absolutely *any* x so what I show has to deal with this. For example, for $\forall x \in \mathbb{R}(x^2 \geq 0)$ I have to show that for any x I am given, then $x^2 \geq 0$.

Example 11.5

We can see the process in action in the earlier example 'For all $x \in \mathbb{Z}$ there exists $y \in \mathbb{Z}$ such that $y > x$.' Here if someone gives me an x I have to find a y such that $y > x$. I can take their x and add 1 to get a bigger integer, that is, I let $y = x + 1$.

Notice in this example that the y I find depends on the given x. This is quite common.

Statements beginning 'there exists'

For statements involving 'there exists', i.e. $\exists x, P(x)$, my goal is to *find* an x that satisfies the condition $P(x)$. For example, to show $\exists x(x^2 + 2x - 3 = 0)$ I need to find a solution. Actually, in general, it may be hard to write down a solution, my point is that it is my *goal* to do so. My hope is that in attempting to do so I shall be able to show one exists even if I have no idea what it is.

For example, it is well known that if $b^2 - 4ac \geq 0$, then there are real number solutions to $ax^2 + bx + c = 0$. I could have applied this knowledge in the example '$\exists x (x^2 + 2x - 3) = 0$' above to say $b^2 - 4ac = 2^2 - 4 \times 1 \times (-3) = 4 + 12 = 16$. Since this is greater than 0 a solution exists. I don't yet know what it is.

Example 11.6

Now consider the statement '$\exists y \in \mathbb{Z}$ such that $\forall x \in \mathbb{Z}, y > x$', which we know is false. We can see this in another way. The statement begins with $\exists y$ so I have to find a y that satisfies the condition that for all x we have $y > x$. So from the start I have to have an integer y. What happens when someone gives me an x equal to $y + 1$ or $y + 1000$? Then we don't have $y > x$, so the statement is false.

Notice the difference in these two examples. In the first, I am given, then I have to find. In the second, I have to find and later I am given.

Example 11.7

If I have to show a set S has an upper bound, i.e. $\exists U \in \mathbb{R}$ such that $\forall s \in S, s \leq U$, then I have to find a U so that if anyone hands me an s from S, then $s \leq U$.

In summary, with 'for all' statements I imagine someone hands me something over which I have no control and for 'there exists' I have to take the initiative and find something amongst the possible x.

In statements involving two quantifiers, this can be played as a two-player game. For example, in '$\forall x \exists y P(x, y)$' I can imagine that someone gives me an x and to win I have to find a y so that $P(x, y)$ is true.

More complicated expressions

Let's look at a more complicated example that uses three quantifiers.

Example 11.8

In Example 11.3 l is the limit of the sequence $a_1, a_2, a_3, \ldots$, if $\forall \epsilon > 0 \, \exists N \geq 1$ such that $\forall n \geq N (-\epsilon < a_n - l < \epsilon)$ is true.

We can read this as the following. If someone brings me an ϵ, then I have to find a natural number N so that whenever someone else brings me an n bigger than N, then the condition $-\epsilon < a_n - l < \epsilon$ is satisfied. True, this is complicated but does give us a handle on how to deal with it, i.e. we know who has to do what. I have to find an N and, since at the start someone gave me an ϵ, the N should probably depend on the ϵ.

This method of seeing the quantifiers as processes is very useful when solving problems.

Example 11.9

Problem: Show that for every pair of natural numbers, there is a natural number greater than both of them.

Solution: Obviously this is true. We can rewrite the statement as $\forall x, y \in \mathbb{N} \, \exists z \in \mathbb{N}(z > x$ and $z > y)$. From this it is obvious that if someone brings to me an x and y I have to find a z. Thus I probably need a proof that says 'Let z equal something in terms of x and y.' It doesn't take long to see that if I set $z = x + y$, then z is bigger than x (as y is a natural number is it is bigger than 1) and z is bigger than y.

Just to reiterate, someone comes along with an x and y, I have to find the z.

Polished solution: Let $z = x + y$. Then $z > x$ as y is positive and $z > y$ as x is positive.

Negation of quantifiers

The negation of statements with quantifiers is remarkably simple.

Suppose that we have the statement 'All cats are grey.' The negation is not 'All cats are not grey.' Note that this does make sense, it means that there are no grey cats. Worse still, it is an everyday response to the statement 'All cats are grey'; imagine someone saying 'No, all cats are *not* grey' to see what I mean.

However, let's look at this logically. If 'All cats are grey' is false, then that means that there is a cat that is non-grey, i.e. there exists a cat which is not grey. So a 'for all' negates to a 'there exists'.

More generally, suppose that the statement $\forall x, P(x)$ is true. The negation of this is that $P(x)$ does not hold for every x. So there must exist at least one x so that $P(x)$ is false, i.e. that not$(P(x))$ is true. In other words, we have the statement $\exists x (\text{not } P(x))$.

From this reasoning and similar reasoning for 'there exists' we can show that

$$\text{not } (\forall x, P(x)) \iff \exists x \, (\text{not } P(x)),$$
$$\text{not } (\exists x, P(x)) \iff \forall x \, (\text{not } P(x)).$$

That is, to negate a statement we change $\forall$ to $\exists$ (and $\exists$ to $\forall$) and negate the sentence after the quantifier.

Examples 11.10

(i)   'For all $x \in \mathbb{Z}$, $x^2 \neq 4$' has negation 'There exists an integer x such that $x^2 = 4$.'
Note that the latter is true. Also note that there is more than one integer with this property.

(ii) 'There exists a mathematician who is not intelligent' has negation 'All mathematicians are intelligent.'

(iii) 'not $(\forall x \exists y\, P(x, y))$' can be written '$\exists x$ not $(\exists y\, P(x, y))$' when we consider '$\exists y\, P(x, y)$' as a statement. By applying the rules again we get '$\exists x\, \forall y$ not $P(x, y)$'.

By arguing as in the last example and since multiple quantifiers are written from left to right we can show the following.

Negation of statements with quantifiers

To negate a statement of the form $Q_1 x_1 Q_2 x_2 \ldots Q_n x_n\, P(x_1, x_2, \ldots, x_n)$, where Q_i is $\forall$ or $\exists$ for $1 \leq i \leq n$, we do the following:

(i)   Change every $\forall$ to $\exists$ and every $\exists$ to $\forall$.
(ii) Replace P by its negation.

This can be shown rigorously using the techniques of Chapter 24. For statements given in words rather than symbols, you may need to tidy up the English a little to make it read well.

Example 11.11

Problem: Show that there is no smallest positive real number.

Solution: This looks like the negation of a statement because it includes the word 'no'. If we use $\mathbb{R}^+$ to denote the positive real numbers, then the statement is actually 'not $(\exists x \in \mathbb{R}^+ \forall y \in \mathbb{R}^+ (x \leq y))$'. Dropping the reference to $\mathbb{R}^+$ and using our procedure above we can rewrite this as $\forall x\, \exists y\, (\text{not}(x \leq y))$. This is the same as $\forall x\, \exists y\, (x > y)$.

This last statement says if someone give me an x, then I need to find a y that is less than it (and is positive of course). That's easy: If I just take y to be half of x (or a third, or a quarter ...), then I get a positive number less than x.

Polished solution: The statement that there is a smallest positive real number is $\exists x\, \forall y$ $(x \leq y)$ so we wish to show not $(\exists x\, \forall y\, (x \leq y))$. This can be rewritten as $\forall x\, \exists y (x > y)$. To show this is true, given x let $y = x/2$, then $0 < y < x$.

Exercises

Exercises 11.12

(i)   Rewrite the following using symbolic notation $\forall$ and $\exists$.
    (a) If a and b are real numbers with $a \neq 0$, then $ax + b = 0$ has a solution.
    (b) If a and b are real numbers with $a \neq 0$, then $ax + b = 0$ has a unique solution.
(ii) Negate the following.
    (a) There exists a grey cat.
    (b) For all cats there exists an owner.

(c) There exists a grey cat for all owners.

(d) Every fire engine is red and every ambulance is white.

(iii) Negate the following:

(a) Some of the students in the class are not here today.

(b) Let $x, y, z \in \mathbb{N}$. For all x there exists y such that $x = y + z$.

(c) There exists unique x such that $P(x)$ is true.

(d) All mathematics students are hardworking.

(e) Only some of the students in the class are here today.

(f) The number $\sqrt{x}$ is rational if x is an integer.

(iv) Simplify the following:

(a) not $(\forall y \, \exists x \, (P(x, y) \implies Q(x, y)))$.

(b) not $(\exists x, y \, \forall z \, \text{not} \, (\forall u \, \exists v \, P(u, v, x, y, z)))$.

(c) not(There exists $x \in \mathbb{R}$ and $y \in \mathbb{R}$ such that for all $z \in \mathbb{Q}$ we have $x \geq z$ and $z \geq y$).

(d) not(There exists $x \in \mathbb{R}$ and $y \in \mathbb{R}$ such that $x \geq y$ or for all $z \in \mathbb{Q}$ we have $x \geq z$ and $z \geq y$).

(v) Show the following:

(a) $\exists N \in \mathbb{N}$ such that $\forall n \geq N, \dfrac{1}{n} < \dfrac{25}{37}$.

(b) $\exists N \in \mathbb{N}$ such that $\forall n \geq N, \dfrac{5n^2 + 2}{n^2} - 5 < \dfrac{1}{1000}$.

(c) $\forall \epsilon > 0, \exists N \in \mathbb{N}$ such that $\forall n \geq N, \dfrac{1}{n} < \epsilon$.

(d) $\forall \epsilon > 0, \exists N \in \mathbb{N}$ such that $\forall n \geq N, \dfrac{5n^2 + 2}{n^2} - 5 < \epsilon$.

Summary

▶ The number of quantifiers is a rough measure of the complexity of a statement.

▶ To show $\forall x \, P(x)$, imagine someone gives you an x and you have to show $P(x)$.

▶ To show $\exists x \, P(x)$, you have to find (any) x that satisfies $P(x)$.

▶ To negate a quantified sentence P, change every $\forall$ to $\exists$ and every $\exists$ into $\forall$ and replace P by its negation.

CHAPTER 12

Examples and counterexamples

Some facts can be seen more clearly by example than by proof.
Leonard Euler

If I were wrong, one would be enough.
Einstein's reply to hearing of the Nazi book *100 Authors Against Einstein*

Now we come to one of the biggest secrets in thinking like a mathematician. So far in this book we have covered many of the important tools for logical thinking: statements, implications, quantifiers. Many other books cover these. But they cover in little detail what I consider to be one of the most important tools. It is fundamental to thinking like a mathematician as I am sure many mathematicians will agree, including those textbook writers. But you won't find this topic given its true prominence in the textbooks.

So what is this great secret tool? Simple! Examples. The quote from Euler[1] is crucial to thinking like a mathematician. First let's remove a possible misconception. Low-level mathematics is often taught in the following way: 'This is how the product rule for differentiation works, here are some examples, now you do some exercises just like the examples.' This is the monkey-see-monkey-do approach. Students are set problems where they can look at the given examples and merely copy the format. (This does actually work well for low-level mathematics.) This type of example is usually called a **worked example**. It is not these I am interested in.

For high-level mathematics the problems are harder (though not in the sense that you have very complicated functions to differentiate). Problems are set that require you to think and to apply what you have learned in situations you have not seen. That's hard!

How do you make it easier? First, stop thinking 'All I need is some more worked examples.' There is no way teachers or books or the world wide web can provide worked examples for every possible question.[2]

---

[1] Pronounced 'Oiler'.

[2] Also, as I like to say to my students, when you get a job and your boss tells you to go and solve a problem, you won't say to them, 'Can you give me some examples of similar problems that someone else has solved?' That's a good way to avoid a successful career.

Worked examples have their place but they do not generate true understanding. So what does? It is the ability to *create* examples.

In this chapter we will look at what is meant by an example, explain why they are a big secret, how they force logical thinking by 'testing' them against statements, and how to create them. An important class of examples we will look at are those known as counterexamples. These are examples we use to show that a statement is false.

Examples

Reversing worked examples

Let us first deal with worked examples. Consider the standard presentation of maxima and minima of functions in the study of calculus. We define first how a function is to be differentiated. Then singular points (also known as turning points) are defined as points where the derivative is zero. Next we are told that there are three types of singular point: maxima, minima and inflection. It is then shown that the second derivative of the function determines the type. After this examples are shown: here's a function, here is where it has singular points, this is the type of singular point. The process is easy, differentiate f, solve $f'(x) = 0$, differentiate to get $f''(x)$ and use the sign of $f''(x)$ to find the type.

This is the standard method of using worked examples. If you learn this method, then given a function you can find the maxima and minima. But what if I turn it around and ask you to create a function f of the variable x with a maximum at $x = 2$ and with a minimum at $x = -6$? This is a far greater test of understanding. It is a lot harder. But in attempting to do it you can learn a lot of mathematics.

Hence, the key ability is to create examples. Often students are given a process and told to answer certain questions using that process. True mathematical thinking comes from creating examples for the reverse of that process.

There are a number of reasons why this is a big secret. The main is that students are not often asked to do it. One reason for this is that lecturers don't like to set questions involving 'create an example' since it causes them work. Such a problem is harder to mark because it has to be checked carefully. It is far easier to look at the bottom of the page at the student's underlined answer for a standard problem. Furthermore, students like to have standard problems because the standard solutions can be easily imitated. Creating good examples is really hard! And it can be dispiriting for the students, which is another reason lecturers don't like to set such problems.

We can summarize the idea as the following.

> Create your own examples: Given a process that finds an answer to a problem, reverse it and ask 'What problem gives this answer'?

For example, we have seen that in calculus we can be given a function and have to determine where the maxima and minima are. We can reverse this by asking for an example of a function that gives predetermined maxima and minima. For instance, the example above of creating a function f of x with a maximum at $x = 2$ and with a minimum at $x = -6$.

Examples of objects

Creating your own examples is an excellent way to learn mathematics but examples have other uses. They are useful for

(i)   testing statements, i.e. testing that we are thinking logically,
(ii)   remembering definitions and statements,
(iii) explanations.

To give you an idea, when presented with a statement you can explore it using examples. By an example, I mean nice simple examples, such as $\mathbb{Z}$ is an example of an infinite set, or $f : \mathbb{R} \to \mathbb{R}$ given by $f(x) = x$ is an example of a function. If someone is talking about a set, then I can use the integers as an example. I can also use the empty set as a good example. Keeping an example in mind while approaching a statement is illuminating, it helps understanding. We shall see this in many of the forthcoming chapters and we shall see in the next section that examples can be used to show that statements are not true.

Examples can be a powerful aid to remembering statements. It is far easier to remember a specific example of some truth and to use it to 'reconstruct' the truth. Non-examples are useful too. When I was first learning analysis I found that a good example for remembering the definition of continuous function was a function that was not continuous. This function identified precisely what was needed in the definition of continuous function.

The third reason above for using examples is that it is easier to explain mathematics, in fact explain almost any subject, with a good example. Thus, when called upon to explain something, drop a good example into the explanation. This impresses the listener (particularly lecturers). It is far easier to latch onto an example than an abstract statement or definition.

Therefore, in summary:

Collect good examples.

What constitutes a good example will be discussed later. At the moment I just want to indicate that examples are important.

Counterexamples

A counterexample is a special type of example.

Definition 12.1

*An example which shows that a statement is false is called a **counterexample**.*

The 'counter' part of the word comes from the fact that we are countering, in the sense of rejecting or rebutting, the truth of a statement.

Examples 12.2

(i)   Consider 'All primes are odd.' The number 2 is a counterexample to this statement.
(ii)   The statement 'If I am Churchill, then I am English' was used as a counterexample to 'A implies B is equivalent to not(A) implies not(B).'

(iii) Consider 'Let p and q be real numbers. If $p/q \in \mathbb{Q}$, then $p \in \mathbb{Q}$ and $q \in \mathbb{Q}$.' This can seem quite reasonable. However, let $p = \pi/3$ and $q = \pi/2$ and we see that $p/q = 2/3$. Thus p and q are real numbers such that $p/q \in \mathbb{Q}$ and yet *both* p and q are not in $\mathbb{Q}$.

Thus $p = \pi/3$ and $q = \pi/2$ provides a counterexample to the statement.

Remark 12.3

A statement of the form 'If ..., then ...' cannot be shown to be true by taking a particular example of where the hypotheses are true and the conclusion is true. However, if we can find just one example for which the statement is not true, then the statement is not a theorem.

Therefore, to show a statement is wrong you need only one counterexample. One and one only is needed.

Exercise 12.4

Find a counterexample to 'The square of a real number is positive.'

How to create examples and counterexamples

A lot of the internal dialogue of a mathematician is about examples and counterexamples, yet they are often neglected in books and lecture courses. As I said earlier it is often very difficult to construct examples and counterexamples, but it is where the true mathematician has a chance to shine.

Thus your first reaction to being given an example is to find another example. But that is not the only time. You should cultivate a sceptical mind. Hence, if someone makes a statement, then your reaction should be to disbelieve them and attempt to find an example that shows the statement is false. Even if their statement is true, the mental workout that this process gives is beneficial. It also helps develop a feel for the statement. Note that constantly doing this in real-life situations can lose you friends – people tend to get upset if you are constantly finding fault with what they say!

> Given any statement try to find a counterexample.

Example 12.5

A letter to a newspaper stated that time travel is impossible because of logic: If time travel were possible, then one would meet lots of people from the future.

Reading this I tried to find many reasons why it was not true. Try it before reading the following.

I had some ideas why it might be wrong. Maybe time travel only allows us to travel forward in time (by amounts larger than we do already!). Maybe time travellers are not allowed to communicate with us. Maybe time travel has a range, you can't travel back more than a year and time travel is still a number of years away (and time travel machines can't be transported).

In the case of mathematics how does one produce these examples? One way is to have a good stock of examples in your head ready to be used at a moment's notice, for example, the empty set or $\mathbb{Z}$.

The other way is practice. That is, the more practice you have in creating examples and counterexamples, then the better and quicker you become at creating them. Just as there is no sure-fire method for problem-solving, there is no magic formula for creating examples. One of the reasons that standard texts avoid examples and counterexamples is that this lack of a magic formula makes it hard to teach their use.

Creating examples and counterexamples is initially frustrating as first attempts are often not very good. It is important to persevere as the ability is key to excellence in mathematics.

Exercises

Exercises 12.6

(i)   Find examples of the following:

(a) A non-constant function $f : \mathbb{R} \to \mathbb{R}$ such that $f(x) = 0$ for a finite number of x.

(b) A non-constant function $f : \mathbb{R} \to \mathbb{R}$ such that $f(x) = 0$ for an infinite number of x.

(c) A non-polynomial function $f : \mathbb{R} \to \mathbb{R}$ such that $f'(x)$ is always positive.

(d) A non-polynomial function $f : \mathbb{R} \to \mathbb{R}$ such that $f'(x)$ is negative for $x < 0$ and positive for $x \geq 0$.

(e) A function $f : \mathbb{R} \to \mathbb{R}$ such that f has a maximum at $x = -2$ and a minimum at $x = 7$.

(ii) Determine whether or not the following are true. Find a counterexample for false statements and show that true ones are true.

(a) $x^3 < 0$ for all $x < -1$.

(b) $x^3 > 0$ for all $x > 1000$.

(c) $x^3 \leq 0$ for all $x \leq 1$.

(iii) Find counterexamples to the following

(a) $(1 - x)/x$ is not an integer for $x \in \mathbb{Z}$,

(b) $(a + b)^2 = a^2 + b^2$,

(c) $\forall x, y \in \mathbb{R}^+ (\sqrt{x + y} = \sqrt{x} + \sqrt{y})$,

(d) $\forall x, y \in \mathbb{R} (\sqrt{x + y} = \sqrt{x} + \sqrt{y})$.

(e) If $x < y + (1/n)$ for all $n \in \mathbb{N}$, then $x < y$.

(f) If $a < b$ and $c < d$, then $ac < bd$.

(g) If $n \in \mathbb{N}$, then $n^2 + n + 41$ is a prime number.

(iv) Find a counterexample to $f'(c) = 0$ implies that f has a maximum or a minimum at c.

(v)   Find previous exercises you have done and decide how they could be 'reversed'. Then do some examples arising from reversing them.

Summary

- ▶ Create your own examples.
- ▶ Collect examples.
- ▶ A counterexample is an example that shows that a statement is false.
- ▶ Only one counterexample is needed to show that a statement is false.
- ▶ Always try to show that a statement is false.

CHAPTER 13

Summary of logic

A summary of ways of writing statement A implies statement B

- A implies B.
- A is true implies B is true.
- $A \Longrightarrow B$.
- If A, then B.
- If A, B.
- B if A.
- A only if B.
- A is sufficient for B.
- B is necessary for A.
- Either A is false or B is true.

Ways of writing statement A is equivalent to statement B

- A is equivalent to B.
- $A \Longleftrightarrow B$.
- A if and only if B.
- A is necessary and sufficient for B.

PART III

Definitions, theorems and proofs

CHAPTER 14

Definitions, theorems and proofs

The highest form of pure thought is in mathematics.

Plato

We now come to the heart of how mathematicians organize and present their work. Look through any high-level mathematics book (or see Chapter 1) and you will find that mathematics is not presented as one continuous piece of prose like a novel. Instead the text is divided up into small nuggets of information such as a Theorem, Proposition, Lemma, Corollary, Proof, Definition and Conjecture. All of these have special meanings and we will see how to approach them in the following chapters.

Meanings

We shall now briefly describe the meanings of the above words.

- Definition: an explanation of the mathematical meaning of a word.
- Theorem: a very important true statement.
- Proposition: a less important but nonetheless interesting true statement.
- Lemma: a true statement used in proving other true statements.
- Corollary: a true statement that is a simple deduction from a theorem or proposition.
- Proof: the explanation of why a statement is true.
- Conjecture: a statement believed to be true, but for which we have no proof.
- Axiom: a basic assumption about a mathematical situation.

Definitions

Much more will be said about definitions in the next chapter. For the moment let us say that in higher mathematics close attention must be paid to definitions – much more than at lower levels. Definitions allow us to separate one class of objects from another or single out some interesting property.

True statements

The words theorem, proposition, lemma and corollary denote statements that are true.

Theorems and propositions

The most important mathematical statements are called **theorems**. Any result of importance will be called a theorem. We use **proposition** for statements that we think are of less importance but which are of some intrinsic interest. It is very difficult to give examples of the distinction between the concepts of theorem and proposition as different authors will put the same statement in different categories. I once saw two distinguished mathematicians argue like children about the difference, but this is rare; most can't be bothered to draw a precise distinction. In fact, in this book we will use theorem to mean any type of true statement.

Examples 14.1

(i)   The following is true: Napoleon was a Corsican.
(ii) Every natural number can be written as a product of primes. (This will be shown in Theorem 25.5.)

Lemmas

A **lemma** is a statement that is a step on the road to proving another statement. Lemmas are considered to be less important than propositions and again the distinction between categories is rather blurred.

An interesting point to note is that often they eventually turn out to be more useful than the statement they are used to prove.

Corollaries

A **corollary** is a statement of interest that is deduced from a theorem or proposition.

Example 14.2

Napoleon was French. (This is true because Corsica was part of France and as we have seen in Example 14.1(i) Napoleon was a Corsican.)

The other terms

Proofs

Mathematicians solve problems – proof is the guarantee that our solutions are correct.

A **proof** is an explanation of why a statement is true. Much, much more will be said about proofs since the defining feature of high-level mathematics is the emphasis on proof. In some sense this book is about proof.

Conjectures

A **conjecture** is a statement which we believe to be true but for which we have no proof. Conjectures are easy to make. Good conjectures are harder to make. A good mathematician will be making, testing and refining conjectures as they work.

Axioms

An **axiom** is a basic assumption about a mathematical situation. Axioms can be considered facts that do not need to be proved (just to get us going in a subject) or they can be used in definitions.

In Euclid's work on geometry he assumed five axioms – for example, between any two points one can draw a line – and from these axioms deduced theorems. The point is that these axioms were the only facts used without proof.

Axioms are also used in definitions. For example, a **group** is a set with some way of 'multiplying' pairs of elements to produce a third element. This multiplication, called a binary operation, has to satisfy some properties; these properties are called axioms. We won't say much about axioms.

Fermat's Last Theorem

It is interesting to consider the case of Fermat's Last Theorem to show just how imprecisely mathematicians use the words above.

Fermat's Last Theorem is famous amongst mathematicians and its story is told in Simon Singh's book *Fermat's Last Theorem*. The statement is the following.

Theorem 14.3 (Fermat's Last Theorem)

There are no positive integer solutions for x, y and z to $x^n + y^n = z^n$ for $n > 2$.

This is perhaps surprising since for $n = 2$, i.e. $x^2 + y^2 = z^2$, there are lots of solutions.

Fermat made this statement in a notebook around 1630 and claimed he had a proof too long for the margin to contain. He never did provide a proof and nor could anyone else until Andrew Wiles gave one in the 1990s. Yet the statement was known as Fermat's Last *Theorem* despite being a *conjecture* for around 350 years. Wiles deduced the result when he proved a much bigger theorem. This theorem is known as the Taniyama–Shimura Conjecture – despite it being a theorem.

Thus, Fermat's Last Theorem, which was a conjecture, is a corollary of the Taniyama–Shimura Conjecture, which is in fact a theorem!

Hopefully, this example should show that mathematicians are not always consistent when using these terms.

Exercises

Exercises 14.4

(i)   Find some books and look at the balance between the number of definitions, lemmas, propositions, theorems and corollaries.

(ii) Find two books where the same statement is called a proposition in one and a theorem in another.

Summary

▶ A definition explains the meaning of a word.

▶ Theorems, propositions, lemmas and corollaries are all true statements.

▶ A proof is the explanation of why a statement is true.

▶ A conjecture is a statement believed to be true, but for which no proof is known.

▶ An axiom is a basic assumption about a mathematical situation.

CHAPTER 15

How to read a definition

In Geometry . . . men begin at settling the significations of their words;
which . . . they call Definitions.
Thomas Hobbes, *Leviathan*, 1651

Precise definitions are vital in high-level mathematics. We need precision so that we can all agree on what we are talking about. Nonetheless, owing to personal preferences, definitions may vary slightly from mathematician to mathematician so it pays to be vigilant.

The reason for giving a mathematical object a particular name is often lost in the mists of time. Frequently names come from ordinary English words but the mathematical meaning may be different to that in everyday speech and give no clue to the definition. For example, the major objects of interest in algebra are groups, fields and rings! Names for objects can be humorous, such as greedoids, or can be derived from a person, such as Gorenstein rings. In the latter case there is no guarantee that the person had anything to do with it – Daniel Gorenstein claimed that he did not know the definition of a Gorenstein ring, despite it being named after him.

What is a definition?

A mathematical definition gives the meaning of a word (or phrase) in a specific way. The word (or phrase) is generally defined in terms of properties.

To begin with let us look at some (rather easy) examples we will need later.

(i)   An integer is **even** if it is the product of 2 and another integer.
(ii)   An integer is **odd** if it is not even.
(iii)   We call a set X with a finite number of elements a **finite set**.
(iv)   If a natural number greater than 1 is divisible only by 1 and itself, then it is called **prime**.
(v)   A prime number p is called a **twin prime** if $p - 2$ or $p + 2$ is prime.
(vi)   A positive integer n is a **square number** if $n = x^2$ for some integer x.
(vii)   A natural number is called **squarey** if its digits are the last digits of its square.
(viii) A prime number is called **squarey-twinney** if it is a twin prime and is squarey.

The last two have been made up to use as examples so do not expect to see them in other books.

The purpose of a definition

The main purpose of a definition is so that everyone knows what we are talking about. For example, we saw in Chapter 1 that some mathematicians define the set of naturals to include 0 and some do not. This is confusing, but even greater confusion would result if the definition being used by an author was different to that of a reader. By making the definition explicit the author avoids this.

In the case of natural numbers two different concepts are being given the same name. We can also have the situation where two different definitions are given but they are in fact the same concept.

The main mathematical reason for giving a definition is to identify some interesting objects worthy of study (although sometimes we give definition to exclude certain 'bad' objects). Also, psychologically, it is easier to deal with a concept once it has been given a name.

Definitions can be used as a solution to a problem. For example, if we define i to be the square root of -1, then we can begin to define complex numbers. Defining complex numbers leads to a good theory of quadratic equations (solutions always exist) and surprisingly helps to solve problems such as ordinary differential equations.

In studying mathematics it is vitally important that you can recall definitions precisely. You have to have all the right conditions or else you are defining something else. A common problem I find amongst students is that they cannot advance in some problem because they do not know the definition of a word. This is an example where students are viewing mathematics as an 'apply-a-process' subject rather than 'understand-the-concepts'.

The 'if and only if' nature of mathematical definitions

Giving definitions is one area where mathematicians are imprecise. In a definition with an 'if', what is intended is an 'if and only if'. The 'only if' part of a definition is considered to be such an obvious part that it is omitted.

For example, the definition above 'A positive integer n is a square number if $n = x^2$ for some integer x' should be read as an 'if and only if' statement:

'A positive integer n is a square number if *and only if* $n = x^2$ for some integer x.'

In other words the number is called square if the condition is true but, more than that, only if the condition is true. There are no square numbers that do not satisfy the condition.

This is an important point to bear in mind when reading mathematical definitions. It is the only time that an 'if' can be read as an 'if and only if'. Do not do this when reading theorems for example.

How to read a definition

Observe

Obviously, given a definition one should observe precisely the conditions given. Bear in mind that we are not allowed to read in anything extra. Note that in an example *all* the conditions need to be true. Not just some.

For example, in the definition of squarey-twinney above, the prime needs to be a twin prime and needs to be squarey. So, for example, if a prime is a twin and not squarey, then it is not squarey-twinney.

What are we dealing with?

The first task is to identify what we are dealing with. Is it something we already know well with an extra property? For example, a twin prime is just a prime p with an additional property: $p - 2$ or $p + 2$ is a prime.

We can ask other questions. Is it similar or different to a definition already known? Is it analogous to something else? Is it a definition we know plus a new condition? For example, a proper subset of X is a subset with the additional property that it is not equal to X.

What examples of this definition exist?

Given a definition, we need to ask if such an object exists. Admittedly, it is unlikely that you will be given a definition of an object that does not exist! The point is to improve understanding by initially being sceptical. If such objects do exist, how common are they? Is the object unique? Is there a finite number? An infinite number?

Returning to our examples beginning on page 103, obviously, even and odd numbers exist. There is obviously an infinite number of finite sets. So, there is a plentiful supply of them, and it is easy to construct examples. What about twin primes? Well, 5 and 7 are twin. So are 41 and 43. But these may be the only ones. If you look at a list of primes less than 1000 you will see that there are more than just these initial examples. Then we can ask, is there an infinite number of twins? Interestingly, nobody knows!

For another example recall squarey numbers above. A natural number is called *squarey* if its digits are the last digits of its square.

Do examples exist for this? Obviously $5^2 = 25$ and 25 ends in a 5, so 5 is squarey. But are there any others? Well, experimentation shows that $76^2 = 5776$, so 76 is also squarey. Looking at the numbers from 1 to 100, we find that 1, 5, 6, 25, and 76 are all squarey. Thus only five out of the first hundred numbers have this property. It is easy to see that this property should in some sense get rarer as our numbers get bigger. This gives us some idea of the notion of squarey.

Find standard examples

From the examples we need to select some standard examples which clearly exhibit the properties and, most importantly, help us to remember the definition. (We have to be careful using them for definitions though as sometimes the example can be misleading. We have to pick the right one!)

For example, when meeting a new definition involving primes, I often use the prime numbers 3, 5 and 13 as examples (the number 2 is a bit extreme – see the next bit of advice) as these are small enough to do calculations with. If I am dealing with infinite sets, then I often take $\mathbb{Z}$, the set of integers.

One of the reasons for finding standard examples is that we can apply them when analysing theorems and deepen our understanding, as we shall see in the next chapter.

Find trivial examples

The concept of being a 'trivial example' is a subjective one – it depends on context. Basically an object is trivial if it is an example in an obvious way. We are looking for very, very simple examples. Trivial examples can help develop a feel for a definition and can be valuable when analysing theorems and their proofs.

The numbers 0 and 1 can often be considered trivial, so when first meeting the definitions above one could ask 'Is 0 an odd or even integer?' (It is even as it is the product of 2 and 0.) For the definition of $n!$ one could ask what 0! should be. (This is explored further in the exercises.)

Similarly, we can ask 'Is 1 a prime number?' In fact, it isn't – numbers had to be greater than 1 in our definition. Why do mathematicians include this condition? Well, we could define 1 to be a prime; certainly it is divisible only by 1 and itself! The problem is that many really good looking statements about primes become false if 1 is a prime. Basically, experience has taught mathematicians that it is not a good idea to have 1 as a prime number.

The empty set $\emptyset$ is a trivial set – you can't get more trivial than a set without elements! We can ask 'Is $\emptyset$ a finite set?' The answer is yes because the empty set has no elements, i.e. zero elements. For finite sets the set with one element is also a useful example that is often trivial.

In Examples 1.12 we see that $X \subseteq X$ and $\emptyset \subseteq X$ are trivial examples of subsets.

For functions to the real numbers or integers we can take $f(x) = 0$, or any other constant function. In geometry, lines, circles and even planes are trivial examples.

It is important to bear in mind that triviality depends on context; in some situations, 0 and 1 are not trivial. (Think binary numbers!)

Find extreme examples

This is similar to the previous piece of advice. Again it depends on context. Here I shall take extreme to mean that the example of the definition is at the boundary of the definition. Hence, one could sometimes argue that trivial objects are extreme. However, often I want something stronger, something pathological. For example, in topology there exist examples of curves that fill the whole plane. We would expect that a curve can cover only a one-dimensional object, but that is not the case. It's an extreme example of what can occur.

For the moment let us consider examples at the boundary of a definition. Thus, in the definition of a subset of X given in Chapter 1, we can consider $\emptyset$ and X as extreme examples since you can't get a smaller subset of X than $\emptyset$ and you can't get a larger subset than X.

For primes, 2 is an extreme example of a prime: It is the first prime and also has the interesting property that it is the only even prime.

Find non-examples

It is a good idea to know some examples that do *not* satisfy the conditions of the definition; these are called **non-examples**. Do not confuse these with counterexamples (which are examples that show a statement is false).

Actually, non-examples are useful for finding counterexamples to statements, but can also be used in fixing the definition in your head. For instance, when recalling the definition of continuous[1] functions I consider a specific non-example, i.e. a discontinuous function.

A non-example of a finite set is the set of integers. The empty set $\emptyset$ is a non-example of an infinite set but is rather extreme, so any other finite set would be better.

Create new objects from old ones

Once a definition is made it is common to make new objects from old ones. For example, in Chapter 1 we saw that we can take subsets, intersection, unions and products of sets.

Furthermore, if the object is a set with a property, then we can ask 'Does the property hold for a subset, intersection, etc.' For example, is a subset of a finite set also finite. Yes it is. Is the subset of a group also a group? No, since the binary operation may not work on the subset. That is, the multiplication of x and y from the subset may not be in the subset. (See the next exercises for a precise definition of binary operation.)

Exercises

Exercises 15.1

(i)   Find examples of the following.

   (a) Let X be a set. Define a **binary operation** on X to be a map $* : X \times X \to X$. That is, $*$ takes two elements of X and produces a third. For two elements x and y we usually write this third as $x * y$. For example, let $X = \mathbb{Z}$ and $* = +$. Then $*$ is just addition on the integers.

   (b) A binary operation is called **commutative** if $x * y = y * x$ for all $x, y \in X$.

   (c) A function $f : \mathbb{R} \to \mathbb{R}$ is called **convex** if for all $x < y$ and $0 \le t \le 1$, we have
   $$f(tx + (1 - t)y) \le tf(x) + (1 - t)f(y).$$

---

[1] The definition of continuous functions is too complicated to go into here – if you study higher mathematics, then you will have met it or will meet it soon.

(d) Let x be a natural number. Take the sum of the squares of the digits to produce a new integer. Repeat this process until it produces a 1 or repeats indefinitely. The integer x is called a **happy number** if the process ends with a 1.

(ii) We all know that, for numbers a, b, c and d, $\dfrac{a}{c} \times \dfrac{b}{d} = \dfrac{a \times b}{c \times d}$ and how useful this identity is. Why can't we define addition of fractions to be something similar? That is, to make calculation easy why don't we define

$$\frac{a}{c} + \frac{b}{d} = \frac{a+b}{c+d}?$$

(iii) The factorial of n, denoted $n!$, was defined earlier as $n! = n \times (n-1) \times (n-2) \times \ldots 2 \times 1$. For all non-negative integers define $n!$ to be the number of ways of arranging the first n numbers (with no repetitions).

Show that the two different definitions coincide for $n \geq 1$ and that $0! = 1$ with the new definition. (We can generalize the definition even further by what is called the Gamma function. Investigate this.)

Summary

▶ Observe the conditions given.
▶ What are we dealing with?
▶ What examples exist?
▶ Find standard examples.
▶ Find trivial examples.
▶ Find extreme examples.
▶ Find non-examples.
▶ Create new objects from old ones.

CHAPTER 16

How to read a theorem

> *The first precept was never to accept a thing as true until I knew it as such without a single doubt.*
>
> René Descartes, *Le Discours de la Méthode*, 1637

Theorems and proofs are central to mathematics. In this chapter I will explain how to deal with a new theorem; consideration of proofs will come later.

As explained earlier, students often conceive of mathematics as being about processes applied to solve set problems, such as using a certain rule to differentiate a product or quotient, or solving a differential equation by first solving a quadratic. Higher-level mathematics moves beyond this, and statements – rather than processes or algorithms – become important.

The use of theorems in solving problems is not obvious. For instance, theorems may not be giving an algorithm. The three theorems we will use in this chapter (given below) tell more about the structure of certain mathematical objects, rather than how to apply an algorithm.

A consequence of this is that students expect theorems to tell them how to do something and so miss what the theorem is really saying.

Note that from now on our default convention will be to call any true statement a theorem and not refer to propositions and lemmas unless necessary.

Three theorems

In order to study theorems we will need some theorems to study. We will use the following three theorems in this chapter.

> **Theorem 1:** If m and n are odd natural numbers, then mn is an odd integer.
> **Theorem 2:** There are infinitely many prime numbers.
> **Theorem 3:** The number $\sqrt{2}$ is irrational, i.e. it cannot be written in the form m/n where m and n are integers.

The first of these you should already be convinced is true and you might like to supply your own proof. The second and third are old theorems (both are over two thousand years old) and are less intuitive. (Both can be proved by contradiction; see Chapter 23.)

Analysing theorems

Find the assumptions and conclusions

As you know from Chapter 7, Implications, theorems can usually be written in the form

'a collection of assumptions imply some conclusion'.

That is, they are in the form, 'If ..., then ...' Identifying precisely what these assumptions and conclusions are is our first goal in dealing with a theorem. There is frequently room for debate but you should be precise in *your* mind what the assumptions and conclusions are.

For the first theorem we have

'm and n are natural and odd' as assumptions

and

'mn is an odd integer' as the conclusion.

We rephrase the second theorem as

'If X is the set of prime numbers, then X is infinite.'

This is an ugly, ugly statement of the theorem – it introduces an unnecessary X for a start – but we can clearly see the assumptions and conclusions.

The third theorem can be rewritten as

'If $x = 2$, then $\sqrt{x}$ is irrational'

or

'If $x = \sqrt{2}$, then x is irrational.'

Is the theorem telling us something about the number 2 or $\sqrt{2}$? The answer is open to debate. Either way, the assumption concerns a specific number and the conclusion is about irrationality.

Rate the strength of the assumptions and conclusions

We want to know how strong the assumptions and conclusions are. The best theorems have weak assumptions and strong conclusions. What constitutes weak and strong is subjective; again there is room for debate.

A **strong assumption** refers to a small set of objects. A **strong conclusion** says something very definite and precise about those objects. In both cases the opposite of strong is **weak**.

Mathematicians want weak assumptions and strong conclusions, that is, we take a very wide collection of objects (weak assumption) and say something very definite about them (strong conclusion).

The assumption of Theorem 2 above is strong; it refers only to the set of prime numbers. (On the other hand if you care only about prime numbers, then this is a weak assumption as it says something about *all* prime numbers. Again, room for debate!)

Similarly, Theorem 3 only discusses properties of the square root of 2 so has a very strong assumption. On the other hand, Theorem 1 has a weaker (or more general) assumption; we are allowed to take *any* odd natural numbers, of which there are many.

We can make the assumptions of Theorem 1 weaker by replacing the natural number assumption with one about integers. This makes the whole statement a *stronger statement* – we want weak assumptions and strong conclusions for a good theorem.

Changing assumptions can reveal how good a theorem is. Many mathematicians do this almost subconsciously when they first meet a theorem.

For Theorem 3 we can weaken the assumption by investigating the square root of prime numbers, say

'If p is a prime, then $\sqrt{p}$ is irrational.'

This statement is true. (Proving it is an exercise in Chapter 23.) It can be weakened further to natural numbers:

'If n is a natural number, then $\sqrt{n}$ is irrational.'

This last statement is *false*, as it is not true for all n, e.g. $\sqrt{4}$ is rational (since it is an integer).

Thus if we weaken the assumptions too much we can get false statements. An important part of mathematical endeavour is to determine how far we can weaken the assumptions. For example, what is the best possible assumption about n so that $\sqrt{n}$ is irrational? Here best means that if we weaken our assumption, then $\sqrt{n}$ is not irrational.

A strong conclusion tells us something definite. All the theorems above have strong conclusions. So, consider the theorem

'Suppose $x^2 + 2x + 1 = 0$. Then $x \geq -1000$.'

This has a very weak conclusion, you could probably guess it was true. The strongest conclusion (that is true) is '$x = -1$'.

More will be said about weakening assumptions in Chapter 33.

Compare with earlier theorems

We can apply our powers of observation. Compare the current theorem with earlier ones. How do the assumptions and conclusions differ? Are they more restrictive, or less restrictive? That is, are they weaker or stronger?

This allows you to place the theorem within the subject as a whole and to see the various inter-relations between theorems.

Observe the detail

In a theorem almost every word will be important – even the little words. Read and notice every word and think about what they mean.

For example, did you notice that in Theorem 1 the assumptions were about *natural* numbers and the conclusion about *integers*?

A good reason for observing is that it is easier to remember what has been explicitly observed. It is not enough to remember the conclusion; the assumptions are of equal importance. Thus, if asked to state Pythagoras' Theorem (well-known and given on page 126), it is not enough just to say '$c^2 = a^2 + b^2$'. You must first explain that a, b and c are lengths of the sides of a triangle and that triangle is in fact right-angled with c the hypotenuse.

Classify what the theorem does and how it can be used

What does a theorem *really* tell us? Does it allow us to calculate, does it classify (i.e. tell us what something is)?

For example, Theorem 1 on the oddness of products could be of use in checking calculations (for example when checking that a computer program for multiplying is working correctly). Theorem 2 tells us something about the structure of the set of primes. If the set of primes was finite, then we could possibly find them all by computer, say, or we could prove theorems about them using this list. As it is, their infinite number suggests that these options are not open to us.

The irrationality of $\sqrt{2}$ statement can be seen as a 'classification' theorem; it classifies what type of number $\sqrt{2}$ is.

Draw a picture

Drawing a picture is always a good idea, though unfeasible at times. It will not help much in the prime numbers or irrationality theorems but works well in suggesting the 'mn-is-odd' theorem is true. Try it!

You should be aware that sometimes pictures can be misleading so be careful.

Apply to trivial examples and other extreme cases

Apply the theorem to trivial and extreme examples of the objects in the assumptions. What happens if a particular number is 0 or 1? What happens if we take the trivial function defined by $f(x) = 0$? What happens if we take the empty set? What happens to the circle or line?

These examples help sharpen understanding. They often give a clearer picture of where the theorem applies.

As an extreme example consider the statement '$y = x^2$ and $z = y^2$, so $z \neq x^2$.' This seems plausible as y and y^2 are generally different, but it is not true. Consider $y = 1$. This occurs when $x = 1$.

Is the converse true?

For any implication given, consider the converse. Recall that the converse of '$A \implies B$' is '$B \implies A$,' and that if '$A \implies B$' is true, then '$B \implies A$' may or may not be true.

Considering the converse forces us to consider the hypothesis and conclusion and sometimes tells us something important.

The converse to our first theorem is

'If mn is an odd integer, then m and n are odd natural numbers.'

This is *not* true. Recall that to show this we need only find a *single* counterexample. Take $m = -1$ and $n = 3$, then the number $mn = -3$ is an odd integer but m is not a natural number.

Lecturers and authors often give the best result possible, so the converse is probably not true, and checking to see if it is appears pointless. But if the converse is not true, then find a counterexample. This will deepen your understanding of the problem and give practice in finding counterexamples, which is very helpful during problem-solving.

Rewrite in symbols or in words

One good way of understanding a theorem written in words is to rewrite it in symbols, and vice versa. Hence, our example of a converse 'If mn is an odd integer, then m and n are odd natural numbers' becomes

$$'mn \in \{x \in \mathbb{Z} \mid x = 2j - 1, \text{ for some } j \in \mathbb{Z}\}$$
$$\implies m, n \in \{x \in \mathbb{N} \mid x = 2j - 1, \text{ for some } j \in \mathbb{N}\}.'$$

Sometimes the theorem is clearer and easier to remember, sometimes – as in this example – it is not.

The conclusion of Pythagoras' Theorem when written as '$c^2 = a^2 + b^2$' is clearer than 'The square of the hypotenuse is equal to the sum of the squares of the other two sides.'

What happens to non-examples?

Consider examples that do not satisfy the assumptions, i.e. non-examples. Does the conclusion still hold true for some of these? Similarly, what happens if one of the assumptions is dropped from the statement of the theorem. It may be easier to answer these questions when the proof is studied.

In Theorem 1 we can consider what happens when only one of m and n is odd and when neither is odd. In these cases we see that the statement is false. Hence, this is a good theorem; we can't weaken the even/odd assumptions any further and produce a true statement. However, we can weaken the assumptions to 'm and n are odd integers' and still produce a true statement. Note that we can't weaken it any further by dropping the even/odd condition.

Generalize

If we drop an assumption from statement A, then we call the weaker statement a **generalization** of A. For example, the statement 'If p is a prime, then $\sqrt{p}$ is irrational' is a generalization of '$\sqrt{2}$ is irrational.'

You should always be looking to generalize statements. If you generalize and can find a counterexample to the general statement, then you will have found that the assumption is vital to the original theorem.

Generalization is very important to thinking like a mathematician and will be dealt with in Chapter 33.

Exercises

Exercises 16.1

(i)   Investigate the following theorems.

(a) Suppose that $m, n \in \mathbb{N}$. Then mn is even if and only if m and n are even.

(b) Suppose that f is convex. Then
$$f\left(\frac{x+y}{2}\right) \leq \frac{f(x)+f(y)}{2}.$$

(c) If A and B are finite sets, then $|A \times B| = |A||B|$.

(d) Let $n \in \mathbb{N}$ with $n \geq 3$. For n distinct points on a circle connect consecutive points by a straight line. The sum of the interior angles of the resulting shape is $(n-2) \times 180°$.

(e) If $n \in \mathbb{N}$ and $n \geq 7$, then
$$\frac{n}{n^2 - 8n + 12} \geq \frac{1}{n}.$$

(f) Let A be a finite set. Let S be the set of all subsets of A. Then $|S| = 2^{|A|}$. (The set S is called the **power set** of A.)

(g) If $f : \mathbb{R} \to \mathbb{R}$ and $g : \mathbb{R} \to \mathbb{R}$ are convex functions, then h given by $h(x) = f(x) + g(x)$ and $m(x) = \max\{f(x), g(x)\}$ are both convex.

(h) For sets A, B and C we have

i. $A \backslash (B \cup C) = (A \backslash B) \cap (A \backslash C)$,

ii. $A \backslash (B \cap C) = (A \backslash B) \cup (A \backslash C)$,

iii. $A \neq B$ if and only if $(A \backslash B) \cup (B \backslash A) \neq \emptyset$,

iv. $A \cup B \subseteq C$ if and only if $A \subseteq C$ and $B \subseteq C$.

What happens in the extreme case(s) where some (or all) sets are empty?

(ii) Analyse the following theorem:

Theorem 16.2

Let a, b, c and d be real numbers that are not all zero. Let

$$\begin{cases} ax + by &= p \\ cx + dy &= q \end{cases}$$

be a pair of equations in the variables x and y with p, q $\in \mathbb{R}$. This system of equations has a unique solution if and only if ab $-$ cd $\neq$ 0.

Summary

- ▶ Find the assumptions and conclusions.
- ▶ Rate the strength of the assumptions and conclusions.
- ▶ Compare with earlier theorems.
- ▶ Observe the detail.
- ▶ Classify what the theorem does and how it can be used.
- ▶ Draw a (useful) picture.
- ▶ Apply the theorem to simple examples.
- ▶ Apply the theorem to trivial and extreme examples.
- ▶ Is the converse true?
- ▶ Rewrite in symbols or in words.
- ▶ What happens to non-examples?
- ▶ Generalize.

CHAPTER 17

Proof

Do not confuse reasons which sound good with good, sound reasons.
Anon.

Mathematics is fantastic. It is a subject where we do not have to take anyone's word or opinion. The truth is not determined by a higher authority who says 'because I say so', or because they saw it in a dream, the pixies at the bottom of their garden told them, or it came from some ancient mystical tradition. The truth is determined and justified with a mathematical proof.

What is a proof?

A proof is an explanation of why a statement is true. More properly it is a *convincing* explanation of why the statement is true. By convincing I mean that it is convincing to a mathematician. (What that means is an important philosophical point which I am not going to get into; my interest is more in practical matters.)

Statements are usually proved by starting with some obvious statements, and proceeding by using small logical steps and applying definitions, axioms and previously established statements until the required statement results.

The mathematician's concept of proof is different to everyday usage. In everyday usage or in court for instance, proof is evidence that something is likely to be true. Mathematicians require more than this. We like to be 100% confident that a statement has been proved. We do not like to be 'almost certain'.

Having said that, how confident can we be that a theorem has been proved? Millions have seen a proof of Pythagoras' Theorem; we can be certain it is true. Proofs of newer results, however, may contain mistakes. I know from my own experience that some proofs given in books and research journals are in fact wrong.

Why prove statements?

In other subjects most statements are open to debate and whether you believe a particular one may be down to personal tastes or prejudices. The existence of proofs means this is not so in mathematics and is a major reason for using them. The great advantage is that you can assess the truth of a statement by studying its proof.

By proving statements we can build mathematics, one statement on top of another. This gives real power to mathematicians. It allows us to be confident and advance. Philosophers, for example, are still arguing about the same questions that the ancient Greeks wrestled with. Not so in mathematics; our subject has moved on greatly from those times.

Another good reason is that writing your own proofs is good for you, it helps expose where your misunderstandings and weaknesses are, and hence lets you know what you need to work on.

Proofs are hard to read

So proofs are good for us as individuals and provide a strong foundation to mathematics, particularly when we come to apply results in real-life situations. And yet students are stubbornly resistant to proofs. I have seen so many students turn away from this wonderful aspect of the subject and ask 'Why do I need to read a proof?' or 'Why do I have to prove things? I'll believe you.' The main complaint is 'I hate proofs. I don't understand them and can't do them.'

I can understand these views. To most students proving statements is new. As noted before, much of pre-proof mathematics is procedural – here's how you solve quadratics, here's how you differentiate a product – proofs just seem to be an extra complication. Also, proofs are hard. They are hard to understand. One reason for this is that proofs are written up so that much of the initial working has been removed. It is that initial working that helped the proof finder find the proof.

The person who gave the original proof did not just sit down and write it out in one go, starting at the start and finishing at the end. Instead, they probably started in the middle, went along to dead-ends, went round in circles, stopped, re-started and generally just haphazardly moved towards the proof. Once they had it they wrote it up, probably going through a number of versions and corrections, leaving out all the false starts and initial guesses.

Yet this path of discovery is not there for the reader. The next chapter attempts to deal with this problem. (And if you want a one-sentence summary of the next chapter, it is: use examples to see the proof.)

As regards the question of whether you should accept things on trust. Why can't you just accept what someone else says is true? Well, you can if you want other people to do your thinking for you, but this book is about thinking for yourself. To really think like a mathematician you must embrace proof.

Proofs are hard to create – but there is hope

Another reason for students not liking proofs is that they are very hard to create. There is no procedure, no algorithm, no road map, or magic procedure for creating proofs. It seems to be an art. Each proof seems to require a unique insight that makes it work. All seems hopeless. All seems lost.

And yet there are many strategies we can apply. We may not all be great artists but our drawings can be improved by knowing about perspective or composition. Mathematics is the same. There are techniques we can employ. For example, let's say we had to prove that an equation has a unique solution. What we do is assume that another exists and then show that their difference is zero, i.e. they are equal. Other simple examples are to find where the hypotheses are used and to consciously look for patterns and similarities in proofs.

The main aim of this book is to give you ideas how to create proofs. Like problem-solving this requires practice. The more proofs you read and study, and the more you write, the better you will become at creating them.

Where we are going

In the next chapter we will see how to pull apart a proof so that you get the most from it. We will then complete Part III with a look at Pythagoras' Theorem and its proof. In Part IV we will look in detail at the following different methods of proof:

- direct,
- cases,
- contradiction,
- mathematical induction,
- contrapositive.

Exercises

Exercise 17.1

Collect together some proofs you have already met.

In particular find proofs which you do not understand. Identify what it is about the proof that you do not understand. Is the start mysterious? Is it at some point in the main body of the proof that you begin to not understand? Is it at the end? That is, you don't understand why the proof is finished.

Summary

- ▶ A proof guarantees that a statement is true.
- ▶ The way a proof is presented is usually very different to how it was first discovered.
- ▶ Proofs are hard to read and hard to create.
- ▶ Reproving statements allows you to develop your own proof-creating skills.
- ▶ To really think like a mathematician you must embrace proof.

CHAPTER 18

How to read a proof

Nullius in verba.
Translation: *Take nobody's word for it.*
Latin proverb

As we have stated, proof is central to mathematics; without proof mathematics loses its power. In this section we show how to approach a proof and break it down to manageable, understandable pieces. Important clues and helpful hints as to why a theorem is true are often removed in the final written version of a proof. The construction lines are erased, and, unfortunately, it is up to the reader to reconstruct them. Think of the proof as being a tight bundle you have to unpack.

You may need some more experience to get the full benefit from the following advice; a lot of it should make sense straight away but rereading at a later date could be useful.

A simple theorem and its proof

The following theorem is easy to understand. It has similarities with Theorem 1 in the previous chapter but is different. The subsequent proof will be the primary example in this chapter.

Theorem 18.1

Let m and n be natural numbers. The product mn is odd if and only if m and n are odd.

Does this theorem seem reasonable? Why is it different to Theorem 1 of the previous chapter? Apply the ideas of that chapter to this new statement.

Proof. Suppose that m and n are odd. Then, by definition, $m = 2k + 1$ and $n = 2j + 1$ for some natural numbers k and j. Then $mn = 2(2kj + j + k) + 1$. Since $2kj + j + k$ is a natural number, say r, we have $mn = 2r + 1$. That is, mn is odd.

Now for the converse. Assume that one of m and n is even. Without loss of generality, we can assume it is m, so $m = 2k$ for some natural number k. We have $mn = 2kn$, that is, mn is divisible by 2 and so is even. Hence, mn is odd only if m and n are not even. That is, mn odd implies that m and n are odd. Hence, the only way mn can be odd is if m and n are odd. □

Note that the end of the proof is marked with a square. This is sometimes called a Halmos tombstone as it was invented by the mathematician Paul Halmos to mark that the proof has been finished.

Don't worry if you don't understand the proof yet, it should become clearer as we apply the ideas below and reconstruct the missing detail. Also, the proof is not correct. It contains a non-fatal error! Can you spot it? (The word non-fatal means that the error can be corrected and the statement proved true.)

How to read a proof

While reading the following advice think about how it applies to the above theorem and proof.

Apply the reading techniques

Apply the reading techniques given earlier. Skim through to gain an overview, identify what is important, ask questions and so on. The aim should be to find the part upon which the whole proof hinges.

Here the proof appears to depend on deductions from calculations.

Break it into pieces

Divide the proof into logically independent sections. Proofs are usually not just one long flowing argument but a number of separate arguments to be analysed, e.g. calculations, verifying statements, etc.

In the above proof one indicator of an independent section is the introductory sentence of the second paragraph informing us that the converse is about to be tackled.

There is another way to see this. The statement of the theorem is an 'if and only if', and so there is the strong possibility – which is the case here – that the proof can be split in two: the 'if' part and the 'only if' part:

(i) 'Let m and n be natural numbers. The product mn is odd if m and n are odd.' (This is the same as 'm and n odd $\implies mn$ odd'.)

(ii) 'Let m and n be natural numbers. The product mn is odd only if m and n are odd.' (This is the same as 'mn odd $\implies m$ and n are odd'.)

Identify the methods used

Mathematicians have a number of methods of proof: calculation, direct, induction, contra-positive, contradiction, cases, counting arguments, etc., some of which are to be detailed in Part IV. Decide which are being used – it is common for a significant proof to be a combination of methods. The 'm and n odd implies mn odd' part of the statement is proved directly through a calculation.

So what about the converse? That is, 'if mn is odd, then m and n are odd'. Certainly, this is not done in a direct method: we do not begin by assuming that mn is odd and proceed with a series of implications to show that m and n are odd. Instead, it is assumed that one of m and n is even. This is the negation of 'm and n is odd'. Thus the contrapositive statement is being proved (recall that 'not $B \implies$ not A' is the same as '$A \implies B$'; see Chapter 8). Notice we are not explicitly told this; it is for us to realize, there is no big announcement 'We will prove the contrapositive statement . . .'.

In more general situations it may be that another theorem is used to prove the new one, or it may the use of a definition, etc. Notice which is used.

Find where the assumptions are used

Identify where the assumptions are used. They will be used once (maybe more) or else they will have been unnecessary. (More will be said about unnecessary assumptions in Chapter 33.) This includes finding where previously proved theorems are used. These will also have assumptions; make sure they are satisfied – be active! An additional point to remember here is that if a theorem gets used again and again in different proofs, it must be important and has the potential to be used in your proofs, so learn it well.

In our proof the assumption that the natural numbers m and n are odd (in the 'if' part of the statement) is used a number of times. In the first paragraph, it is used in setting $m = 2k + 1$, etc., since odd numbers are of this form. The identification of the assumptions for the 'only if' is a bit harder since we assume the negation of a statement so we can apply the contrapositive method – it is a bit disguised but it is used nevertheless.

Apply the proof to an example

A very effective method to understand a proof is to apply each step to a particular special or concrete case that satisfies the assumptions. This is an important nugget of information to take away from this book.

In the above example this is largely trivial, yet we can have a go. Suppose that $m = 3$ and $n = 7$ and rewrite the proof using those figures. Of course for the proof of the 'only if' part you need to assume that m is even, so let's say it is 6. No assumption is made on n being odd or even, try both. Again, this a matter of being active!

This method can work particularly well for problematic proofs, but unfortunately doesn't shed light in every case.

Draw a picture

Reading a proof is like solving a problem so apply the techniques from Chapter 5, How to solve problems, such as drawing a picture.

In the above proof the picture would look like the following.

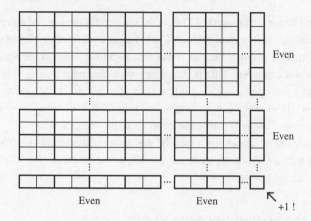

Figure 18.1 A picture to clarify the proof of Theorem 18.1

The picture should convince you. The rectangles all have an even number of squares, except the single square by itself. Hence there is an odd number of squares.

A warning is in order here. We should be careful with pictures, they can be exceedingly misleading. Many mathematicians have been fooled by the power of a picture.

Check the text

Once again, be active. Check that the text is correct. Check that the theorems used do apply. No theorems are used in our example, but if a proof says 'by Theorem 6.3', make sure Theorem 6.3 can be applied. If a proof says 'a calculation shows', do that calculation.

In our example the assertion that $mn = 2(2kj + j + k) + 1$ is not immediately obvious and so we should work it out in full:

$$
\begin{aligned}
mn &= (2k+1)(2j+1) \\
&= 4kj + 2j + 2k + 1 \\
&= 2(2kj + j + k) + 1.
\end{aligned}
$$

In the example proof above we see in the second paragraph that it is assumed 'without loss of generality' (see page 250 for a fuller discussion) that m is even. Let's check. Suppose that it wasn't even. That means n is even as the assumption is that (at least) one of them is even. So simply relabel the numbers, with m as n and n as m. Hence, there is no loss of generality.

Look for mistakes – try extreme cases

Mathematicians are human and make mistakes. Actively look for these mistakes. Be sceptical of everything and really try to show the text is wrong. Search for hidden assumptions and check that the theorems used really apply.

Try extreme cases, such as extreme functions like the zero function. In the above try m equals 1, this is the first odd natural number, so is in some sense extreme. Then $m = 2k + 1$

gives $1 = 2k + 1$, that is $k = 0$. Aha! This is not a natural number as claimed in the proof, it does not come from $\{1, 2, 3, 4, \dots\}$. Hence, the proof is wrong! However, it is not fatally flawed and we can repair it by either including 0 in our definition of natural numbers, or saying k is in $\mathbb{N} \cup \{0\}$, or k is an integer. Either way, the statement of the theorem is true, the original proof was in error, albeit slightly.

A good extreme case to take is when a number is zero as the following shows. Consider the statement:

'Theorem' 18.1

Suppose that a, b, c and d are natural numbers. If $ab = cd$ and $a = c$, then $b = d$.

Proof. We have

$$ab = cd$$
$$\Longleftrightarrow ab = ad, \text{ as } a = c,$$
$$\Longleftrightarrow \ b = d, \text{ by cancellation.}$$

$\square$

This might look convincing. The problem comes when we consider the extreme case of $a = c = 0$. We would then have $ab = cd$, regardless of what b and d were, so say $b = 4$ and $d = 3$, then we have $0 \times 4 = 0 \times 3$, but $4 \neq 3$. This shows how important it is to check a statement with extreme cases.

The faulty reasoning above can also be applied to sets: it is possible to show $A \times B = C \times D$, and $A = C$ implies $B = D$. Here, to avoid problems, we need to assume that A is not the empty set – which is the analogue of zero in this case.

Another extreme case is when two of the 'inputs' in an assumption are equal. For example, I have seen one of Euclid's geometry axioms stated as 'Given x and y in the plane there exists a unique line joining them.' Seems reasonable: we can see the picture in our heads. But what if the two points chosen are in fact the same point? Then there is an infinite number of lines through the point. The correct statement of the axiom is 'Given *distinct* x and y in the plane there exists a unique line joining them.'

Apply to a non-example

Identify precisely where a generalization to non-examples would fail. This can be a very effective learning tool.

For example, the proof that $\sqrt{2}$ is irrational is given in Chapter 23. We could use the proof to attempt to prove that $\sqrt{4}$ is irrational. This statement is not true but merely trying to use the proof can show how and why the proof works for $\sqrt{2}$.

Reflection

As always, reflect afterwards. Is the proof like any other proof you have seen? Some types of proof appear again and again; realizing this makes them easier to learn and use in your work.

How to memorize a proof

The easiest way to remember a proof is to understand it. Learning something by memorizing or rote learning is the hard way of doing it – seek to understand. On the other hand, at some point you may have to memorize, say to improve speed during an exam.

Know how a proof is constructed and know the key point of every paragraph. Is it to do a calculation? Is it where the key assumption is used? Is it setting up notation? These are easier to remember and record. What type of proof is it? Which techniques from Part IV does it use?

Attempt to sum up the proof in a few sentences and try to record what it means – not what it says. So for the above, record

'Use $m = 2k + 1, n = 2j + 1$, a calculation gives the product is of the form $2r + 1$. Converse: use contrapositive. A without loss of generality argument gives mn even.'

Notice that I don't say 'assume m is even'. That is because, if we are using the negation of m and n odd, then we automatically conclude one of m and n is even. You should be well versed in working out negations.

Exercises

Exercises 18.2

(i)   Analyse the collected proofs from Exercise 17.1 using the techniques of this chapter. Do the proofs become clearer? At least a little?

(ii) Analyse this proof of Theorem 16.2 from the exercises of Chapter 16.

Proof. Labelling the equations (1) and (2) in the obvious way, then $d(1) - b(2)$ gives the equation $(ad - bc)x = pd - qb$ and hence we have $x = (pd - qb)/(ad - bc)$. Similarly we can show that $y = (qa - pc)/(ad - bc)$. This solution is obviously unique.

For the converse, as a, b, c and d are not all zero, we can assume without loss of generality that a is non-zero. Hence, $x = (p - by)/a$ is a solution to (1) for every $y \in \mathbb{R}$. Now, assuming (1) has this solution, and because $ad - bc = 0$ we have

$$cx + dy = q \iff cx + \frac{bc}{a}y = q \iff cax + bcy = aq$$
$$\iff c(ax + by) = aq \iff cp = aq.$$

Hence, if $cp = aq$, then a solution to (1) gives a solution to (2), and therefore the system has an infinite number of solutions.

On the other hand, if $cp \neq aq$, then a solution to (1) is not a solution to (2), and therefore no solutions exist. □

(iii) Consider the statement, also from Chapter 16 exercises: For sets A and B we have $A \neq B$ if and only if $(A \backslash B) \cup (B \backslash A) \neq \emptyset$.

Proof. Suppose that $A = B$. Then

$$(A \backslash B) \cup (B \backslash A) = (A \backslash A) \cup (A \backslash A) = \emptyset \cup \emptyset = \emptyset.$$

If $A \neq B$, then there exists $x \in A$ but $x \notin B$ or there exists $x \in B$ but $x \notin A$. In the former we have $x \in A \setminus B$ hence $x \in (A \setminus B) \cup (B \setminus A)$. Therefore $(A \setminus B) \cup (B \setminus A) \neq \emptyset$. A similar reasoning proves the result in the latter possibility.

$\square$

Analyse the proof (and don't forget to consider the extreme case of empty sets!).

Summary

- ▶ Apply the reading techniques.
- ▶ Break the proof into pieces.
- ▶ Identify the methods used.
- ▶ Find where the assumptions are used.
- ▶ Apply the proof to an example.
- ▶ Draw a picture.
- ▶ Check the text.
- ▶ Look for mistakes.
- ▶ Apply the proof to a non-example.
- ▶ Memorize by understanding.

CHAPTER 19

A study of Pythagoras' Theorem

Reason is immortal, all else mortal.
Pythagoras, *Diogenes Laertius (Lives of Eminent Philosophers)*

Pythagoras' Theorem is probably the best-known mathematical theorem. Even most non-mathematicians have some vague idea that it involves triangles and squaring something known as the hypotenuse.

Because the ideas in the previous chapters on 'How to read a theorem' and 'How to read a proof' are so important we will apply them to this famous theorem to see them in action. So in this chapter we will pull apart the theorem and its proof, we'll see a converse for it and also a generalization.

Statement of Pythagoras' Theorem

As you are a budding mathematician, you probably have a better idea than a non-mathematician of what the statement is, but here it is again.

Theorem 19.1

For a right-angled triangle the square of the hypotenuse is equal to the sum of the squares of the other sides.

Exercise 19.2

Use the ideas from Chapter 16, How to read a theorem, to analyse the theorem. Compare your analysis with the one given below.

Study of the theorem

We now analyse the theorem as though we were meeting it for the first time. Obviously we would check what all the words mean, for example, what is a hypotenuse? This is fairly obvious, but what about the other techniques in Chapter 16? We shall apply them now.

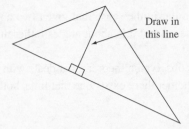

Draw in
this line

Figure 19.1 Any triangle gives two right-angled triangles

Find the assumptions and conclusions

The statement given for Pythagoras' Theorem is a good example of a statement not in the form '$A \Longrightarrow B$' or 'if..., then...'. We can rewrite it in this form in a number of ways. For example,

'If T is a right-angled triangle with sides a, b and hypotenuse c, then $c^2 = a^2 + b^2$.'

This make it obvious that the assumptions concern all right-angled triangles: 'T is a right-angled triangle with sides a, b and hypotenuse c' and that the conclusion is an equation relating the lengths: '$c^2 = a^2 + b^2$'.

Rate the strength of the assumptions and conclusions

Let's rate the assumptions and conclusions. The assumption is about right-angled triangles. Certainly, there are many examples of these, but they are only a small subset of *all* triangles. Thus we may be tempted to say that this is quite weak but not too weak. But consider this. For any triangle we can produce two right-angled triangles; see Figure 19.1. Thus, despite initial impressions, this theorem will tell us something about all triangles. This means it is a very weak assumption. That is good. The more examples a theorem applies to the better.

The conclusion is a very concrete and definite statement. It is an equation. If we just had an inequality (say $c^2 \geq a^2 + b^2$), then the conclusion would be less impressive. This is a strong conclusion.

Note also that the conclusion allows us to calculate: given the lengths of any two sides of a right-angled triangle we can calculate the third. Being able to calculate is always good for a conclusion.

From this we know we have a great theorem which will be very useful as it allows us to calculate attributes of a large number of objects. In fact, as you may know, it allows us to calculate distances in Cartesian coordinates.

Compare with previous theorems

We don't have many theorems to compare with as this is a very simple theorem from the foundations of geometry. Another theorem from the foundations is that in a triangle (not necessarily right-angled) the angles add up to 180 degrees (or 2π radians if you like).

We can see this is similar in the sense that given two angles we can calculate the third. It is different in the sense that it is about angles rather than about lengths. Thus they are quite complementary!

Now that we have observed a theorem that deals with lengths and one that deals with angles we can ask whether there exists one that links both. Indeed there is one.

Exercise 19.3

Find a theorem that relates angles and lengths in a triangle.

Observe the detail

There are not many details to observe in Pythagoras' Theorem as it is rather simple. The detail to remember perhaps is that the triangle is right-angled – not, for example, isosceles. (See the exercises at the end of the chapter for some infamous examples of mistakes with the statement.)

Classify what the theorem does and how it can be used

We have already discussed above what the theorem does: it allows us to calculate the length of a side of a triangle given the lengths of the other two.

Draw a picture

Since this is a geometric theorem – it is about triangles – then we have good reason to draw a picture. In this case draw lots of triangles and measure the lengths. Do the lengths satisfy the equation? They should do but remember we can only measure lengths approximately.

The classic picture is the $(3, 4, 5)$-triangle given in Figure 19.2. This is often mistakenly given as a proof of the theorem. However, it is only a single example, not a proof.

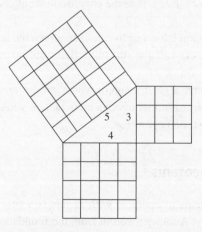

Figure 19.2 The classic $(3, 4, 5)$-triangle

Apply the theorem to simple examples

In drawing the pictures we have already applied the theorem to simple examples such as the $(3, 4, 5)$-triangle.

Apply the theorem to trivial and extreme examples

The assumptions concern right-angled triangles. What are trivial and extreme examples of such objects? I would say that the trivial examples are when $a = 1$ and $b = 1$ so $c = \sqrt{2}$. This is a triangle with two sides of rational lengths and one side of irrational length. One could view this as a counterexample to the statement 'If two sides of a triangle are rational, then so is the third.'

Another useful example is $a = 1$, $b = 2$, and so $c = \sqrt{5}$. This just shows us what happens when one side is twice the length of the other. Another interesting example is $a = 1, b = \sqrt{2}$, and so $c = \sqrt{3}$. I like this, it involves the first three numbers 1, 2 and 3. The triangle $a = 1, b = \sqrt{3}$ and $c = 4$ is good as it has an angle of $\pi/6$.

Now for extreme examples. What is an extreme right-angled triangle? We can view $a = b$ as an extreme triangle because it is a very special case of a right-angled triangle. Here $c^2 = 2a^2$, so $c = \sqrt{2}a$. This means that if a is rational, then c cannot be rational (since a rational number times an irrational number, in this case $\sqrt{2}$, is irrational).

Another extreme we can go to is to make one of the sides very big or very small compared with the other. If we let b tend to zero, then c^2 must tend to a^2 as b^2 gets small. In other words c tends to a. We can see this in a diagram. Just draw a triangle with a very small b as compared with a; you can see that a and c are almost equal. Thus we have shown that the algebra and the geometry are linked in the way we would expect. The algebra agrees with the geometry.

Note that if b equals zero, then we do not have a triangle!

Apply the theorem to non-examples

Let us consider non-examples of the theorem. Here non-examples are triangles that are not right-angled. However, in this case how do we even define hypotenuse? This is always defined as the length opposite the right-angle. There is no way of identifying a special length, so there is no way to identify which should go on the right-hand side of $c^2 = a^2 + b^2$. We need another method of identifying c. Maybe we have to take the longest side. This is reasonable since, for $c^2 = a^2 + b^2$ to be true, we must have $c > a$ and $c > b$. To see this, consider that $b^2 > 0$ is true, thus $c^2 > a^2$, and so $c > a$. Similarly $c > b$.

If you try drawing a few examples, then you will probably see that the equation does not hold, even taking into account approximations in measuring. This should lead us to ask 'Are there *any* non-right-angled triangles for which $c^2 = a^2 + b^2$?' We shall deal with this in the next section when we ask 'Is the converse true?'

We shall also look at non-examples in Exercises 19.12.

Rewrite in symbols or words

We have already written the theorem in as many symbols as possible when we gave the 'if …, then …' statement. Note that is not sufficient to say only '$c^2 = a^2 + b^2$'. We have to explain what a, b and c are, and that c is the hypotenuse in a right-angled triangle.

Proof of Pythagoras' Theorem

As mentioned earlier, it is not enough to show the theorem works approximately for some examples or even perfectly in some special cases such as the $(3, 4, 5)$-triangle. We need to prove that it holds for all triangles. Maybe you were told that the theorem was true by some authority figure in the past and that is good enough for you. However, a central aim of this book is to encourage you to think for yourself and that involves checking any argument given to you.

We will now see a proof of the theorem. There are literally hundreds of proofs.[1] My favourite proof is geometrical.

Proof (of Pythagoras' Theorem). The proof can be shown using the two squares in Figure 19.3. To draw the first square begin by drawing a general triangle with sides a and b and then extend these edges by lengths b and a respectively. Then we can complete the drawing to get the square on the left-hand side of Figure 19.3.

We can draw another square like the one on the right-hand side of the figure. From the figure we can see that both squares have equal area and so we can conclude that

$$\text{Area of left square} = \text{Area of right square}$$
$$c^2 + (4 \times \text{Area of } (a, b)\text{-triangle}) = a^2 + b^2 + (4 \times \text{Area of } (a, b)\text{-triangle})$$
$$c^2 = a^2 + b^2.$$

$\square$

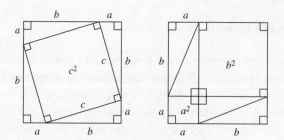

Figure 19.3 Proof of Pythagoras' Theorem

[1] There are over 360 in Elisha Loomis, *The Pythagorean Proposition*, National Council of Teachers of Mathematics, 1968.

Exercise 19.4

Use the ideas from Chapter 18, How to read a proof, to analyse the proof. Compare your analysis with the one given below.

Not all the suggestions in Chapter 18 are relevant. Let us try some of them to help pull the proof apart.

Find where the assumptions are used

Where have we used the assumption that the triangle is right-angled? It is used in constructing the first square. If sides a and b did not meet at a right-angle, then we could not construct a square. And hence could not conclude that the area was the same as the square on the right.

Check the text

If you check the text, then you may see that one fact has been used but not explicitly stated. The picture is very convincing in that the area labelled c^2 in the left-hand picture certainly looks square. But notice that we have not proved that it is square. We know that all the edges are the same length but this does not mean that the shape is a square – think of a diamond or kite shape.

How do we know it is truly a square? Well, we need to show that one of the internal angles is right-angled, once we get it for one, the same proof will work for all. (Can you see why?) Call the internal angle γ.

Suppose that the triangle in Figure 19.4 has small angle α and larger angle β. Then we can see that $\alpha + \beta + \gamma = 180°$ since we have a straight line:

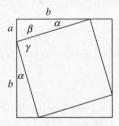

Figure 19.4

But we know that $\alpha + \beta + 90° = 180°$ since the angles in a triangle add up to $180°$ and our triangle is right-angled. Using these two equations we deduce that γ is $90°$ as required.

If we add this explanation, then the argument will be even more convincing. True, the picture looks convincing already, but we have to be careful with pictures – they can mislead.

Exercise 19.5

How else could the picture fool us?

What about the converse?

Let's now look at an important technique for exploring theorems: is the converse true? To do this it helps to write the statement in an 'If ..., then ...' form:

If T is a right-angled triangle with sides a, b and hypotenuse c, then $c^2 = a^2 + b^2$.

The converse will be

If $c^2 = a^2 + b^2$, then T is a right-angled triangle with sides a, b and hypotenuse c.

Note that this does not quite make sense because in the assumptions we do not know what a, b and c are. Let's rewrite Pythagoras' Theorem as the following:

Let T be a triangle with sides of length a, b and c with c the longest. If T is a right-angled triangle, then $c^2 = a^2 + b^2$.

Now the converse becomes:

Theorem 19.6 (Converse of Pythagoras' Theorem)

Let T be a triangle with sides of length a, b and c with c the longest. If $c^2 = a^2 + b^2$, then T is a right-angled triangle.

Is this true or not? It is! But why?

Proof. Let T be a triangle with sides of length a, b and c where C is the angle opposite to the side of length c, then the Cosine Rule states that

$$c^2 = a^2 + b^2 + 2ab \cos C.$$

Now suppose that $c^2 = a^2 + b^2$, then we have $2ab \cos C = 0$. As a and b cannot be zero we must have $\cos C = 0$. This implies that $C = 90° + 180°n$, where n is some integer. Since $0 < C < 180°$ we must have $C = 90°$, i.e. T is a right-angled triangle. $\qquad\square$

Exercise 19.7

Apply the methods of Chapter 18, How to read a proof, to the above proof.

Remark 19.8

The proof of the converse used the Cosine Rule, the proof of which uses Pythagoras' Theorem (see page 22 and following). Note that the Cosine Rule is more general than Pythagoras' Theorem, since we can deduce Pythagoras from it: if we have a right-angled triangle, then $C = 90°$ and so $c^2 = a^2 + b^2 - 2ab \times 0 = a^2 + b^2$.

Thus, we can see that a more general statement can use the simpler theorem in its proof.

Since Pythagoras' Theorem and its converse are true we have two equivalent statements (i.e. If $A \implies B$ and $B \implies A$, then $A \iff B$). This means we can state the following theorem.

Theorem 19.9

Let T be a triangle with sides of length a, b and c with c the longest. Then T is a right-angled triangle if and only if $c^2 = a^2 + b^2$.

Proof. We can prove this by combining the proof of Pythagoras we gave and then using the converse argument we just gave. Note that we can't just use the Cosine Rule in both directions since Pythagoras' Theorem is used in the proof of the Cosine Rule! Did you spot that?   □

A bit more on understanding a converse

Suppose that we have a triangle with sides of length 2.0, 2.1 and 2.9. We have $2.0^2 + 2.1^2 = 2.9^2$. The triangle is therefore right-angled.

Exercise 19.10

Did we use Pythagoras' Theorem or its converse to deduce this?

Suppose that we have a triangle of sides of length 3.6, 7.7 and 8.4. In this case, $3.6^2 + 7.7^2 = 72.25 \neq 70.56 = 8.4^2$. Thus the triangle is not right angled.

Exercise 19.11

Did we use Pythagoras' Theorem or its converse to deduce this?

A common mistake in answering this is to make an argument that the sides do not satisfy the equation $c^2 = a^2 + b^2$ so this can't use Pythagoras' Theorem. Therefore, it must use the converse. This is wrong.

   The correct argument is that Pythagoras' Theorem says that if you have a right-angled triangle, then it sides satisfy the equation. Thus if the equation is *not* satisfied, then there is no way that the triangle could be right-angled. In effect, we are using the contrapositive statement to Pythagoras' Theorem, i.e. an equivalent statement: If $c^2 \neq a^2 + b^2$, then T is not right-angled.

Exercises

Exercises 19.12

(i)   Pythagoras' Theorem is often misquoted. In the film *The Wizard of Oz*, the Scarecrow is given a diploma and to show how clever he has become he points his finger to his temple and says 'The square root of the hypotenuse is equal to the sum of the square roots of the other sides for an isosceles triangle.'

   This is most definitely not Pythagoras' Theorem as it involves square roots and isosceles triangles. However, just because it is not Pythagoras' Theorem does not mean that the statement is false! Apply our methods to analyse this statement. Is it true? If not, then give a counterexample.

(ii)   The theorem is also misquoted in the long-running animated comedy *The Simpsons*. In the episode *$pringfield (or, how I learned to stop worrying and love legalized gambling)* Homer finds some glasses and puts them on and à la Scarecrow puts his finger to his temple and states 'The sum of the square roots of any two sides of an isosceles triangle is equal to the square root of the remaining side.' Another character then shouts 'That's a right-angled triangle, you idiot!'

Why are Homer and the other character *both* wrong?

(iii)   Many theorems from lower-level mathematics are given without proof. Find as many examples of this as you can. Try to find proofs for them from other sources and analyse these using the methods of Chapter 18, How to read a proof. Some examples you may like to try:

(a) The sum of angles in a triangle is 180 degrees.

(b) Subtracting a negative is equivalent to adding a positive.

(c) The value of π is $3.14159\ldots$

(d) The definition of sine and cosine don't depend on the triangle. (You will need to know about similar triangles.)

(e) For any angle θ, $\sin^2\theta + \cos^2\theta = 1$.

(f) The area of a circle of radius r is πr^2.

(iv)   Consider the triangles with sides of the following lengths. Decide which are right-angled and state whether you used Pythagoras' Theorem or its converse.

(a) 7, 24, 25,   (b) 28, 45, 52,

(c) 36.9, 80.0, 88.0,   (d) 0.8, 1, 4, 1.7.

(v)   Construct a right-angled triangle such that the hypotenuse has irrational length but the other two sides have rational length.

(vi)   Construct a right-angled triangle such that the hypotenuse has rational length but the other two sides have irrational length.

(vii) Let X and Y be distinct points in a plane. Draw a line between them. At the midpoint draw a line perpendicular to the line. See Figure 19.5. This line is called the **perpendicular bisector** of X and Y.

Show that for every point p on the perpendicular bisector the distance from X to p and the distance from Y to p are the same. We say p is **equidistant** from X and Y.

(viii) Go back to previous chapters and exercises and re-analyse the theorems and proofs. Did you observe facts that you did not observe before?

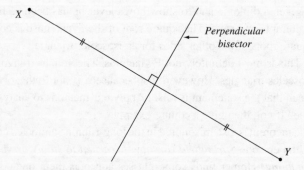

Figure 19.5 Perpendicular bisector

Summary

- ▶ Pythagoras' Theorem is: Let T be a triangle with sides of length a, b and c with c the longest. If T is a right-angled triangle, then $c^2 = a^2 + b^2$.
- ▶ The converse of Pythagoras' Theorem is: Let T be a triangle with sides of length a, b and c with c the longest. If $c^2 = a^2 + b^2$, then T is a right-angled triangle.

Techniques of proof

CHAPTER 20

Techniques of proof I: Direct method

> *Sometimes the simple approach is the best.*
> Popular saying

Our first method for proving theorems is the most straightforward and is called the direct method. Most statements can be broken down into smaller statements of the form 'If A, then B.' To prove A implies B directly you prove A implies A_1, prove A_1 implies A_2, A_2 implies A_3 and so on until you get A_r implies B. Then you have proved $A \implies B$. Each implication should be in some sense obvious, i.e. we should have no giant leaps that are hard to understand!

Examples of the direct method

We have already seen an example of the direct method in the proof of the 'if' part of Theorem 18.1. In this case we have a simple calculation to prove the statement. Another example is the following.

Theorem 20.1

Let m be an integer. If m is odd, then m^2 is odd.

Proof. If m is odd, then $m = 2r + 1$ for some integer r. Then,

$$m^2 = (2r+1)^2 = (2r+1)(2r+1) = 4r^2 + 2r + 2r + 1 = 4r^2 + 4r + 1$$
$$= 2(2r^2 + 2r) + 1.$$

That is, m^2 is odd. $\qquad\qquad\square$

Recall the definition of cardinality from Chapter 1 and examine the next theorem which is a solution to Problem (ii) on page 42 of Chapter 5.

Theorem 20.2

Let A and B be finite sets. Then $|A \cup B| = |A| + |B| - |A \cap B|$.

Apply the methods of Chapter 18 to the above. What happens to trivial examples, say $A = \emptyset$? What about extreme cases, such as $A = B$? What about non-examples, for instance, what happens when one or both of the sets are infinite? Why is the theorem restricted to finite sets? What are the assumptions and how good are they? (Very good, since we only assume that the sets are finite.) How good is the conclusion? (Very good, since it gives us an equation that allows us to calculate. Ways to calculate are always useful!) Can we draw a picture? Yes we can because we can draw a Venn diagram.

Now let's look at the proof.

Proof. If we count all the elements in A and then all the elements in B, then we will count the elements that are in A or are in B, but will count those that are in both A and B twice. In the union of A and B, the elements that are in both are only counted once. Thus $|A| + |B| - |A \cap B|$ counts elements that are in A or in B only once. Hence this is equal to $|A \cup B|$. $\square$

Let's try another simple (admittedly not particularly useful) example involving more steps and this time let us see the thinking behind how we might prove it.

Theorem 20.3

Suppose that $p \in \mathbb{Q}$ and $p^2 \in \mathbb{Z}$. Then, $p \in \mathbb{Z}$.

What are our assumptions? One is $p \in \mathbb{Q}$. Well, we don't yet know many theorems about rational numbers, all we really know is the definition, i.e. we know $p = a/b$ for some $a, b \in \mathbb{Z}$, and we can assume that this is in its lowest form.

The next assumption is that $p^2 \in \mathbb{Z}$. This means that $(a/b)^2 \in \mathbb{Z}$, i.e. $a^2/b^2 \in \mathbb{Z}$. This fraction is also in its lowest form. But this can only happen if $b = \pm 1$ because $a^2 \in \mathbb{Z}$ (as $a \in \mathbb{Z}$). So we can see that $p = a/(\pm 1) = \pm a \in \mathbb{Z}$.

Let's write a polished version of the proof:

Polished solution:
Proof. By assumption $p = a/b$ for some integers a and b, where the fraction is in its lowest form. Thus $p^2 = (a/b)^2 = a^2/b^2$. Since $p^2 \in \mathbb{Z}$ and the fraction is also in its lowest form we have that $b^2 = 1$. Thus $b = \pm 1$ and we can deduce that $p = a/(\pm 1) = \pm a \in \mathbb{Z}$. $\square$

Notice that we used the definition of a rational number, i.e. $p = a/b$. This is a good idea to apply when trying to find a proof of a statement.

Theorem 20.4

Let m and n be real numbers. If $n > m > 0$, then

$$\frac{m+1}{n+1} > \frac{m}{n}.$$

One way to prove this would be to take the right-hand side to the left-hand side and prove something is bigger than 0. Instead we shall assume what had to be proved and work from there. This is a perfectly reasonable method of problem-solving. It is not reasonable to use in the polished version, as we shall see later.

We have

$$\frac{m+1}{n+1} > \frac{m}{n}$$

$(m+1)n > m(n+1)$ (note, as m and n are positive, $>$ does not reverse)

$$mn + n > mn + m$$

$$n > m.$$

Excellent! We have just proved $(m+1)/(n+1) > m/n$ implies $n > m$ and along the way we used m and n positive. This is of course *not* what had to be proved but we might have some hope that the argument reverses. And indeed it does:

Polished solution:

Proof. We have

$$n > m$$
$$\implies mn + n > mn + m$$
$$\implies (m+1)n > m(n+1)$$
$$\implies \frac{m+1}{n+1} > \frac{m}{n}, \text{ as } m \text{ and } n \text{ are positive.}$$

$\square$

Note that our proof used the assumptions, i.e. $n > m$ is used immediately and $n > 0$ and $m > 0$ are used at the end. Also note that the final proof is very different to how we found the proof. Consider the cleaned-up version; the second line is that $mn + n > mn + m$. If we had started the initial investigation with $n > m$ would we have seen that adding mn to both sides is the way to go? It isn't obvious!

In the next example we again use the technique of assuming what had to be proved. It is important to note that we only assume this to find the proof, we do not use this assumption when we write the final proof.

Theorem 20.5

The product of two real negative numbers is positive.

Let us assume that we don't know how to do this. I'm allowed to use facts like $0 \times a = 0$ and the distributive law, i.e. $x(y + z) = xy + xz$.

We want $(-x) \times (-y) > 0$ if x and y are positive. We can consider

$$(-x) \times (-y) = (-1) \times x \times (-1) \times y = (-1) \times (-1) \times xy.$$

So really we need to prove that $(-1) \times (-1)$ is positive since this argument reverses.

We've now reduced to showing that $(-1) \times (-1) = 1$ using simple laws of algebra. Right, let's explore this equation. If we assume it is true, what happens? I stress again that we are assuming that something is true to explore it. We should not assume it to be true when we want to write the proof.

The only thing to do is to put everything on the same side:

$$(-1) \times (-1) = 1$$

$$(-1) \times (-1) - 1 = 0.$$

We note now that we can gather terms, i.e. the -1s:

$$(-1) \times (-1) = 1$$
$$(-1) \times (-1) - 1 = 0$$
$$(-1) \times (-1) + (-1) = 0$$
$$(-1)((-1) + 1) = 0$$
$$(-1) \times 0 = 0.$$

We have arrived at something that is true. Have we proved that $(-1) \times (-1) = 1$? Not really. We assumed our statement was true and produced something that was true. We need to go in the other direction: Start with something that is true and work towards our statement.

The obvious action is to reverse this argument! I can use $(-1) \times 0 = 0$ to prove that $(-1) \times (-1) = 1$.

Let's do this properly.

> Polished solution:
> **Proof.** We have
>
> $$(-1) \times 0 = 0$$
> $$\implies \quad (-1)((-1) + 1) = 0$$
> $$\implies \quad (-1) \times (-1) + (-1) = 0$$
> $$\implies \quad (-1) \times (-1) + (-1) + 1 = 1$$
> $$\implies \quad (-1) \times (-1) + 0 = 1$$
> $$\implies \quad (-1) \times (-1) = 1.$$
>
> $\square$

Exercises 20.6

(i) Write a polished version of the corollary of $(-1) \times (-1) = 1$ that the product of two negative numbers is positive. That is, give the full proof of Theorem 20.5.

(ii) Which rules of algebra have we implicitly used in our proof? Do any of them rely on the product of two negatives being positive? If it did, then we would have what is known as a **circular argument**. This is where we assume a statement is true and later use its truth to prove it. For example, 'I am the best candidate for President because I am better than all the other candidates' is a circular argument. It basically says 'I am the best because I am the best.'

How to show that an equation holds

There are a number of methods for showing that an equation holds. The following is for me the most important.

> To show that an equation holds it is generally better to choose the more complicated side of the equation and make substitutions to reduce that expression to the other side.

If you start with the equation and attempt to rearrange it, then you run the risk of going in circles.

For example, to prove that $\tan x + \cot x = 2\operatorname{cosec} 2x$ for all $x \in \mathbb{R}$ such that $x \neq \dfrac{n\pi}{2}$ for $n \in \mathbb{Z}$, we do the following:

$$\begin{aligned}
\tan x + \cot x &= \frac{\sin x}{\cos x} + \frac{\cos x}{\sin x}, \text{ by definition of tan and cot,} \\
&= \frac{\sin^2 x + \cos^2 x}{\cos x \sin x} \\
&= \frac{1}{\cos x \sin x}, \text{ using } \sin^2 x + \cos^2 x = 1, \\
&= \frac{1}{(1/2)\sin 2x}, \text{ using a half angle formula,} \\
&= 2\operatorname{cosec} 2x, \text{ by definition of cosec.}
\end{aligned}$$

Other options are available for showing that $x = y$:

(i) $x - y = 0 \iff x = y$,
(ii) $x \leq y$ and $x \geq y \iff x = y$,
(iii) $x = z$ and $y = z \iff x = y$.

The first of these is approached in a way similar to the example above. We take $x - y$ and attempt to reduce it to 0. The second will be exemplified in the proof of Theorem 27.20. The third is fairly self-explanatory; we show that x and y are equal to some other object z. This can be used to show that the recurring decimal $0.99999\ldots$ is equal to 1. First we know that $1 = 3/3$. Also we know $3/3 = 3 \times (1/3)$ but $1/3$ is well known to have decimal expansion $0.33333\ldots$ and so $3 \times (1/3) = 3 \times 0.33333\cdots = 0.999999\ldots$. So, as $1 = 3/3$ and $0.99999\cdots = 3/3$, we have their equality. This result is fairly surprising to most people. We intuitively feel that as 1 and $0.9999\ldots$ are distinct decimal expansions they should represent different numbers. This shows that this intuition is wrong.

If and only if proofs

The statement of Theorem 18.1 of 'How to read a proof' was of the 'if and only if' type. This was proved by breaking the proof into an 'if' part and an 'only if' part. This is a common method of attack for proving such statements. In the following example we shall see this in action and also that – sometimes – the proof in one direction reverses to give the other direction.

Theorem 20.7

Let $a \neq 0$, b, and c be real numbers. Then

$$ax^2 + bx + c = 0 \iff x = \frac{-b \pm \sqrt{b^2 - 4ac}}{2a}.$$

The 'only if' part (the $\Rightarrow$ direction) says that if we have a quadratic equation, then it has solutions of the form $x = \dfrac{-b \pm \sqrt{b^2 - 4ac}}{2a}$. But note that it doesn't say that an x of that form *is* a solution – the 'if' part (the $\Leftarrow$ direction) does that.

So let us attempt to prove the problem by dealing with each direction separately. The latter of these, the $\Leftarrow$, is easy to prove. Like most equations, when you have the answer you can just substitute it in easily.

Exercise 20.8

Write out the proof for the 'if' part ($\Leftarrow$ direction).

We now consider the $\Rightarrow$ direction, the 'only if' part of the statement. First, by dividing by a, we can reduce to solving $x^2 + (b/a)x + (c/a) = 0$. Thus without loss of generality we may assume that we are solving $x^2 + \alpha x + \beta = 0$. Later we can substitute back in a, b and c using $\alpha = b/a$ and $\beta = c/a$.

We can complete the square to solve a quadratic of this form. Thus we have the following

$$
\begin{aligned}
& x^2 + \alpha x + \beta = 0 \\
\Longrightarrow \quad & \left(x + \frac{\alpha}{2}\right)^2 - \left(\frac{\alpha}{2}\right)^2 + \beta = 0 \\
\Longrightarrow \quad & \left(x + \frac{\alpha}{2}\right)^2 = \left(\frac{\alpha}{2}\right)^2 - \beta \\
\Longrightarrow \quad & x + \frac{\alpha}{2} = \pm\sqrt{\left(\frac{\alpha}{2}\right)^2 - \beta} \\
\Longrightarrow \quad & x = -\frac{\alpha}{2} \pm \sqrt{\left(\frac{\alpha}{2}\right)^2 - \beta} \\
\Longrightarrow \quad & x = -\frac{b}{2a} \pm \sqrt{\left(\frac{b}{2a}\right)^2 - \frac{c}{a}} \\
\Longrightarrow \quad & x = -\frac{b}{2a} \pm \sqrt{\frac{b^2}{4a^2} - \frac{ac}{a^2}} \\
\Longrightarrow \quad & x = \frac{-b \pm \sqrt{b^2 - 4ac}}{2a}.
\end{aligned}
$$

So now we should be happy because we have proved the result both ways: $\Rightarrow$ and $\Leftarrow$. We can write it up in two parts.

However, let us reflect. We should go over our proof again. There is in fact a better way. We have proved one implication by using a long list of implications. We should ask 'Can all the implications be reversed and still be true?'

The answer is yes! Check it for yourself. So in our answer we would not do the 'if' and the 'only if' separately, we can miss out our earlier attempt at $\Leftarrow$, we just write the $\Rightarrow$ part and can change all the $\Rightarrow$ to $\Longleftrightarrow$.

Notice also that if you wished to memorize the proof, then memorizing each step would be the hard way of doing it. Instead, notice the main elements of the proof:

(i)   It is $\Longleftrightarrow$ all the way.
(ii) We reduce the equation to the simpler case of having 1 as the coefficient of x^2, i.e. we divide by a.

(iii) We complete the square.

(iv) Make substitutions to reverse the reduction made in (ii).

That's it. The rest is just reducing expressions into a simpler form.

Proving that one set is a subset of another

If we have to prove that one set is a subset of another, then we can use the definition, which is $X \subseteq Y$ if (and, of course, only if) $x \in X$ implies that $x \in Y$ for all x.

Theorem 20.9

Let A and B be sets. Then $A \cap B \subseteq A \cup B$.

Proof. Suppose $x \in A \cap B$. Then,

$$x \in A \cap B$$
$$\implies x \in A \text{ and } x \in B, \text{ by definition of } \cap,$$
$$\implies x \in A, \text{ by definition of logical and,}$$
$$\implies x \in A \text{ or } x \in B, \text{ by definition of logical or,}$$
$$\implies x \in A \cup B, \text{ by definition of } \cup.$$

$\square$

Note that we take a general element of the proposed subset and see where that leads. This is a common method of attack.

Exercise 20.10

Give a counterexample to the statement 'For all sets A and B, we have $A \cap B \subset A \cup B$.'

Proving that two sets are equal

We have seen how to show that one set is a subset of another. We shall now see how to prove that two sets are equal.

An obvious way to do this to have a set of equalities linking the two.

Example 20.11

Let A, B and C be sets. Then,

$$A \cap (B \cup C) = (A \cap B) \cup (A \cap C).$$

Proof. We have

$$A \cap (B \cup C) = \{x \mid x \in A \cap (B \cup C)\}$$
$$= \{x \mid x \in A \text{ and } x \in B \cup C\}$$
$$= \{x \mid x \in A \text{ and } (x \in B \text{ or } x \in C)\}$$
$$= \{x \mid (x \in A \text{ and } x \in B) \text{ or } (x \in A \text{ and } x \in C)\}$$
$$= \{x \mid x \in A \cap B \text{ or } x \in A \cap C\}$$
$$= \{x \mid x \in (A \cap B) \cup (A \cap C)\}$$
$$= (A \cap B) \cup (A \cap C).$$

$\square$

Another method utilizes the following proposition, which is similar to the idea of showing that, for two numbers x and y, $x = y$ if and only if $x \leq y$ and $x \geq y$.

Proposition 20.12

Let X and Y be sets. Then, $X = Y$ if and only if $X \subseteq Y$ and $Y \subseteq X$.

Proof. If $X = Y$, then by definition X and Y have exactly the same elements. So if $x \in X$, then $x \in Y$. But simply by definition of a subset this says that $X \subseteq Y$. Similarly if $y \in Y$, then $y \in X$, so $X \subseteq Y$.

Now, for the converse, assume that $X \subseteq Y$ and $Y \subseteq X$. If $X \subseteq Y$, then every element of X is in Y. If $Y \subseteq X$, then every element of Y is in X. This means X and Y have the same elements, i.e. $X = Y$. $\square$

Exercise 20.13

Analyse the proof using Chapter 18, How to read a proof. Can you improve the proof?

Thus we reduce the proof of showing two sets are equal to showing each is a subset of the other. This is a very powerful idea and appears regularly in mathematics so put it in your toolbox.

Exercises

Exercises 20.14

(i)   Use the methods of Chapter 18, How to read a proof, to find the error in this direct proof: If n is even, then $2^n - 1$ is not prime.
'Proof': We have

$$n \text{ even } \Rightarrow n = 2k \text{ for some } k \in \mathbb{N}$$
$$\Rightarrow 2^n - 1 = 2^{2k} - 1 = \left(2^k\right)^2 - 1 = (2^k - 1)(2^k + 1).$$

Thus, $2^n - 1$ can be written as the product of two numbers, so is not a prime.

(ii) Show that the sum of two consecutive odd numbers is a multiple of 4. What is the converse and is it true?

(iii) Prove that for all integers x with final digit equal to 5, x is a multiple of 5. Is the converse true?

(iv) Let $f : X \to Y$ be a function between two sets X and Y. Suppose A and B are subsets of X. Show the following:

(a) $f(A \cup B) = f(A) \cup f(B)$,

(b) $f(A \cap B) \subseteq f(A) \cap f(B)$.

Give a counterexample to the statement $f(A \cap B) = f(A) \cap f(B)$ and construct an example to show that this equality can be true sometimes.

(v) Recall the definition of complement of a set from Chapter 1. Let A and B be subsets of X. Prove or disprove the following.

(a) $(A^c)^c = A$.

(b) $(A \cup B)^c = A^c \cap B^c$.

(c) $A \cap B = A \backslash (A \cap B^c)$.

(d) $(A \cup B) \backslash A = B \backslash A$.

(vi) Prove that

(a) $A \cap (B \cup C) \subseteq B \cup (A \cap C)$,

(b) $(A \times B) \cap (C \times D) = (A \cap C) \times (B \cap D)$,

(c) $(A \times B) \cup (C \times D) \subseteq (A \cup C) \times (B \cup D)$. Are the sets in fact equal?

(vii) Suppose x and y are rational numbers such that $x < y$. Prove that there exists $z \in \mathbb{Q}$ such that $x < z < y$. That is, between every two rationals there exists another rational.

Recall from Chapter 11, Complexity and negation of quantifiers, that to show something exists we should try to construct it from what we know, i.e. construct z from x and y.

(viii) Prove that the following statements are equivalent for A and B subsets of X.

(a) $A \subseteq B$,

(b) $A \cap B^c = \emptyset$,

(c) $A^c \cup B = X$.

(ix) Prove that the following statements are equivalent.

(a) There exists unique x such that $P(x)$ is true.

(b) $\exists x \, (P(x) \text{ and } \forall y (P(y) \Longrightarrow y = x))$.

(c) There exists x so that for all y $P(y)$ is true if and only if $y = x$.

(d) $\exists x \, P(x) \text{ and } \forall y, z((P(y) \text{ and } P(z)) \Longrightarrow y = z)$.

(x) Let $x_1, x_2, x_3, \ldots x_n$ be an arrangement of the numbers $1, 2, 3, \ldots n$. (For example if $n = 6$ we could have $4, 3, 6, 1, 2, 5$.) Prove that if n is odd, then $(x_1 - 1)(x_2 - 2)(x_3 - 3) \ldots (x_n - n)$ is even. (Hint: Consider sums.)

(xi) Let x be an integer. Prove that x^n is odd if and only if x is odd. (Hint: One direction uses the contrapositive and the other uses the binomial theorem.)

Summary

▶ In the direct method, to show $A \Longrightarrow B$ we assume A and proceed to show B.

▶ We can prove that $x = y$ via one of the following:

    (i)   Choose the side with the most complicated expression and reduce it to the other side;

    (ii) $x - y = 0$;

    (iii) $x \leq y$ and $x \geq y$;

    (iv) $x = z$ and $y = z$.

▶ To prove $X \subseteq Y$: assume $x \in X$ and proceed to show that $x \in Y$.

▶ To prove that two sets X and Y are equal:

    (i)   use a string of equalities, or

    (ii) prove that $X \subseteq Y$ and $Y \subseteq X$.

▶ To prove that '$A \Longleftrightarrow B$' show '$A \Longrightarrow B$' and '$B \Longrightarrow A$'.

▶ Summary of summary:

    (i) $x = y$ is equivalent to $x - y = 0$,

    (ii) $x = y$ is equivalent to $x \leq y$ and $y \leq x$,

    (iii) $X = Y$ is equivalent to $Y \subseteq X$ and $Y \subseteq X$,

    (iv) '$A \Longleftrightarrow B$' is equivalent to '$A \Longrightarrow B$' and '$B \Longrightarrow A$'.

CHAPTER 21

Some common mistakes

To err is human. To really mess up takes a computer.
Anon.

The direct method of proof is probably the most basic and will be used in the other methods. Despite this there are a couple of pitfalls that are easy to fall into. Two of the most common are assuming what had to be proved and incorrect use of equivalence. We shall investigate these in this chapter. And since it would be nice to gather common mistakes together in one handy chapter rather than having them distributed throughout the book some other mistakes are included. Also brought in is an explanation of why we can't divide by zero – a mistake you probably already know about but may not have been given a reason why. I have seen all these errors made and, like most mathematicians, have made them myself.

Don't assume what had to be proved

Probably the most common mistake in proofs is assuming what had to be proved. Suppose that we had to prove statement P. If we assume it is true, then it is not surprising that we can deduce it is true; $P \implies P$ would seem to be very obviously true. Another error in this vein is that P is assumed to be true and this is used to deduce something that is true and so it is concluded that P is true. This is of course an incorrect argument. (It says $P \implies Q$, but Q is true, so P is true.)

As an example, consider the following statement:

'If a and b are real numbers, then $a^2 + b^2 \geq 2ab$.'

A fallacious proof is:

'We have

$$a^2 + b^2 \geq 2ab \implies a^2 - 2ab + b^2 \geq 0 \implies (a - b)^2 \geq 0.$$

The last inequality is true as the square of a number is always non-negative, so $a^2 + b^2 \geq 2ab$.'

The error here is that the conclusion (in the statement to be proved) has been assumed (i.e. that $a^2 + b^2 \geq 2ab$) and has led to something we know is true. However, we cannot conclude that a statement is true just because it implies a known truth.

Consider the true statement '$-1 = 1 \Longrightarrow 1 = 1$' discussed on page 65. The '$-1 = 1$' part implies '$1 = 1$', which is true but obviously we cannot conclude from '$1 = 1$' that '$-1 = 1$'.

The real proof is just a reverse of the argument; begin with $(a - b)^2 \geq 0$, something we know is true, and proceed to the conclusion we want by reversing the implication signs above. Try it and see!

It is ok to assume what has to be proved when finding a proof!

Now for the really confusing piece of advice. I have just said that you should not assume what had to be proved when proving statements. However, when it comes to solving problems, i.e. *finding* the proof, it is ok to assume what had to be proved. We saw examples of this repeatedly in the previous chapter.

This strategy will often unlock the problem and allow us to create a proof. In the above example we saw how assuming the conclusion $a^2 + b^2 \geq 2ab$ led us to $(a - b)^2 \geq 0$, something we know is true. Fortunately, in this case – it might not be true in other cases – we can reverse the implication arrows and go from $(a - b)^2 \geq 0$ to $a^2 + b^2 \geq 2ab$.

Thus, when solving problems it is ok as an exploratory tool to assume what had to be proved, just to see where it leads, but when writing a polished version it is not ok.

Square root is a function so it gives a single number

Strictly speaking this is not a mistake in logic but because many students are unaware of what the square root sign really means it does lead to errors in logic.

It is quite common to be taught that the square root of a number gives two numbers: a positive and negative one. This is not actually true. If $\sqrt{}$ is a function, then, by definition of a function, we should only get one value out for each input. So $\sqrt{4} = 2$ and not $\sqrt{4} = \pm 2$. Therefore, to solve $x^2 - 4 = 0$ we say 'rearrange to get $x = \pm\sqrt{4}$ and hence $x = 2$ or $x = -2$' and not '$x = \sqrt{4}$, hence $x = 2$ or $x = -2$.'

On the other hand, if square root is not a function, then what is it? And, from a practical perspective, if it is not a function, then how are we going to differentiate it? Also, if we really are wedded to the idea that two numbers should come out, then why do we write the $\pm$ symbol in the formula for solving quadratic equations $x = \dfrac{b \pm \sqrt{b^2 - 4ac}}{2a}$ from the previous chapter? Since if we decided the square root function gave two values, then it would be sufficient to write $x = \dfrac{b + \sqrt{b^2 - 4ac}}{2a}$ as this would give two solutions.

In summary, always take square root of a to be the positive solution to the equation $x^2 = a$. (And note that the solutions to $x^2 = a$ are $x = \pm\sqrt{a}$.)

Another problem with square roots

Square roots cause a lot of problems. This is partly psychological (because of the above) and partly that in exercises we often remove square roots to get closer to an answer. A common mistake is exemplified by the following. If we have $\sqrt{x^2 + y^2} - 2x = 4$, then we cannot just square every term to get $x^2 + y^2 + 4x^2 = 16$. To see why, consider that what is being asserted is $a + b = c \implies a^2 + b^2 = c^2$. We can easily give a counterexample to this.

Get the implications right

When solving equations we pass from one line to the next by showing that the first implies the second. However, it is important to check that the second implies the first. What we want is a sequence of equivalences to find all the solutions and no more.

The following equation and its solution should make clear this point:

$$\sqrt{x + 3} = x + 1$$
$$x + 3 = (x + 1)^2, \text{ squaring both sides,}$$
$$x + 3 = x^2 + 2x + 1$$
$$0 = x^2 + x - 2$$
$$0 = (x - 1)(x + 2).$$

So $x = 1$ or $x = -2$ are our solutions. Substituting $x = 1$ into the original equation shows we have a correct solution. But if we substitute $x = -2$ into the equation we get $\sqrt{-2 + 3} = -2 + 1$, i.e. $\sqrt{1} = -1$. Thus $x = -2$ is not a solution (we are using the square root operation correctly!). The problem arises because if we have an equation $f(x) = g(x)$ then squaring both sides and solving gives solutions to $-f(x) = g(x)$ as well.

Precision in our use of implications signs will clarify the solution:

$$\sqrt{x + 3} = x + 1$$
$$\implies \quad x + 3 = (x + 1)^2, \text{ squaring both sides,}$$
$$\iff \quad x + 3 = x^2 + 2x + 1$$
$$\iff \quad 0 = x^2 + x - 2$$
$$\iff \quad 0 = (x - 1)(x + 2).$$

The first $\implies$ cannot be reversed (because $a^2 = b^2$ does not imply $a = b$ but that $a = \pm b$) and so at the end we know that we might have too many solutions.

This also shows how important it is to use the 'implies that' sign correctly. This has been mentioned on page 37.

Don't divide by zero (or why 1/0 is not ∞)

The next common mistake is also not really a mistake in logic. However, we can use logic to show why we shouldn't make this mistake.

You may have seen a proof of $1 = 2$ to exemplify that we should not divide by zero. Such an operation is easily missed so be vigilant!

One argument is as follows: Let $a = b$. Then,

$$ab = a^2, \text{ since } a = b,$$

$$ab - b^2 = a^2 - b^2, \text{ by subtracting } b^2 \text{ from both sides,}$$

$$b(a - b) = (a + b)(a - b), \text{ by factoring,}$$

$$b = a + b, \text{ by dividing both sides by } a - b,$$

$$b = 2b, \text{ as } a = b,$$

$$1 = 2.$$

The division by zero happens when we divide by $a - b$. Since we assumed that $a = b$ this must be zero. That is where the argument goes wrong.

This throws up the question 'Why can't we divide by zero?' Let's answer a different question. Suppose we *could* divide by 0. Then 'What do we get when we divide by zero?' For example, what is $1/0$? We could argue as follows:

> Look at $1/x$ and let x get smaller and smaller. One can see that as x gets smaller, then $1/x$ gets larger. Therefore we could *define* $1/0$ to be 'infinity'.

Mathematicians denote infinity by ∞, so we are claiming that $1/0 = \infty$. That at first seems fine. After all, we can see that $2/0$ must be $2 \times (1/0) = 2 \times \infty = \infty$ – it makes sense that 2 times infinity is infinity, doesn't it?

The problem comes when we consider what $0 \times \dfrac{1}{0}$ is. Zero times anything should be zero, so

$$0 \times \frac{1}{0} = 0 \times \infty = 0$$

since if we have zero infinities, we are saying we have no infinities, so the answer must be zero.

On the other hand, the rules of arithmetic are that we can cancel: $a \times (b/a) = b$ so we must have

$$0 \times \frac{1}{0} = 1$$

Thus, combining these two results we get

$$0 = 0 \times \frac{1}{0} = 1.$$

In other words, by allowing division by zero we produce an obviously false equality, i.e. $0 = 1$.

The point is that if we allow division by zero we cannot have both that $0 \times x = 0$ for all x and that $a \times (b/a) = b$ for all a and b. Unfortunate, but true. So if $x/0$ is given a value – even if the value is ∞ – we quickly end up in a tangle.

As mathematicians we can choose the rules we want. Not all choices lead to a useful theory. Indeed, you can construct a theory that defines $1/0$ as ∞ but then you have to lose useful rules like $a \times (b/a) = b$.

In the case of infinity we should treat it not as a number, and so subject to algebraic rules, but as a concept. More will be said on infinity in Chapter 30.

The number $-x$ is not always negative!

Here's an interesting psychological mistake. Let x be an integer. It is surprising how many people will a few lines later use the 'fact' that $-x$ is negative.

It is easy to see that this is false. Take any negative integer, say -3, then $-x = -(-3) = 3$, which is positive.

The problem is that psychologically we see the minus sign and our brains automatically assume that the number must be negative.[1]

So $-x < 0$ is not necessarily true!

Negatives and inequalities

The natural psychological assumption that all undefined numbers are positive causes problems when dealing with inequalities.

For example, if $1/x < 5$, then it seems natural to rearrange to $1/5 < x$. When dealing with equalities we often move expressions from one side to the other without thinking. For inequalities we have to be careful. Take $x = -2$ in the above. Then $1/x = 1/(-2) < 5$ is true, but then $1/5 < -2 = x$ is not. The rearrangement we did earlier was in error.

The error arises because we use the idea that we can multiply both sides by x. In an inequality multiplying both sides by a negative number will reverse the inequality.

Thus in the above $1/x < 5$ becomes $1/5 < x$ if $x > 0$ and becomes $1/5 > x$ if $x < 0$. This process of breaking into cases will be dealt with in the next chapter.

Exercises

Exercises 21.1

(i)   What is wrong with the following argument? 'We have

$$2 = 4 \implies 2\pi = 4\pi \implies \sin 2\pi = \sin 4\pi \implies 0 = 0.$$

Therefore $2 = 4$.' Which of the implications does not reverse?

(ii)   Consider the theorem: For all $n \in \mathbb{N}$, the number $n^2 + 5n + 6$ is not prime. Find the mistake in the following 'proofs':

(a)   For $n = 2$ we have $n^2 + 5n + 6 = 2^2 + 10 + 6 = 20$ which is not prime. Hence the theorem is true.

(b)   Suppose that n is a natural number, so in particular $n > 0$. If $n^2 + 5n + 6$ is not prime, then $n^2 + 5n + 6 = pq$ for some natural numbers p and q such that $0 < p < n^2 + 5n + 6$ and $0 < q < n^2 + 5n + 6$. Since $n^2 + 5n + 6 = pq$ and p and q do not equal 1 or $n^2 + 5n + 6$, then $n^2 + 5n + 6$ is not prime, which was to be shown.

---

[1] Also, children are often taught to pronounce $-x$ as 'negative x'.

(iii) Find the error in the following proof that the sum of two rational numbers is rational. Suppose that m and n are rational numbers, then we can write $m = p/q$ and $n = r/s$ where p, q, r, s are integers (and q and r are non-zero). Then

$$m + n = \frac{p}{q} + \frac{r}{s}.$$

Since the sum of two fractions is a fraction, then $m + n$ must be a fraction. Hence, $m + n$ is rational.

(iv) Above is a proof that we cannot define $1/0$ to be ∞. However, maybe we can deduce it from some other fact!

We can all agree that $1/\infty = 0$ since if we look at $1/a$, then as a goes to ∞, then $1/a$ goes to 0. Now suppose that $1/0 = x$ for some x. Then, $x = 1/(1/\infty)$. Since $1/(1/y) = y$ for all y we must have $x = \infty$. That is, $1/0 = \infty$.

Find *all* the errors in this argument.

(v) Here's another error that people make. What is it?

The following proves the statement $\exists x \in \mathbb{R} \, \forall y \in \mathbb{R} \left((y+1)x^2 = \dfrac{1+y}{y^4} \right)$.

Suppose that $x = 1/y^2$, then

$$(y+1)x^2 = (y+1)\left(\frac{1}{y^2}\right)^2 = \frac{y+1}{y^4} = \frac{1+y}{y^4}.$$

(vi) There are fancy names for common mistakes in mathematical logic. Find out what 'Denying the hypothesis' and 'Begging the question' are.

Summary

▶ Don't assume what had to be proved.
▶ Square root is a function.
▶ The square root of a is the positive solution to the equation $x^2 = a$.
▶ $1/0$ is not ∞.
▶ $-x$ may not be negative.
▶ Be careful with inequalities and negative numbers.

CHAPTER 22

Techniques of proof II: Proof by cases

Little by little does the trick.

Aesop, *The Crow and the Water Jar*

We have already seen that $x = y$ can be proved by showing that $x \leq y$ and $x \geq y$. In other words we have broken the problem into two cases. This is a very common procedure in mathematics. Often we can break down some problem and tackle each case individually utilizing different methods, or even the same methods, in each case.

The famous **Four Colour Problem** is that given a map (of the geographical kind) we need only four colours to colour the map so that no two countries bordering each other have the same colour.[1] The proof of this involved using theory to cut down the problem to a finite number of types of maps. As there were only about one thousand such types a computer was be used to check that all could be coloured with four colours.

Examples of cases

Example 22.1

The number $n^2 + 3n + 7$ is odd for all $n \in \mathbb{Z}$.

We can divide this into two cases: (i) n is even and (ii) n is odd.

If n is even, then $n = 2k$ for some integer k by definition of even. Thus

$$
\begin{aligned}
n^2 + 3n + 7 &= (2k)^2 + 3(2k) + 7 \\
&= 4k^2 + 6k + 7 \\
&= 2(2k^2 + 3k + 3) + 1.
\end{aligned}
$$

Hence the expression $n^2 + 3n + 7$ is odd when n is even.

Now for the other case. If n is odd, then $n = 2k + 1$ for some integer k. We have

$$
\begin{aligned}
n^2 + 3n + 7 &= (2k + 1)^2 + 3(2k + 1) + 7 \\
&= 4k^2 + 10k + 11 \\
&= 2(2k^2 + 5k + 5) + 1.
\end{aligned}
$$

---

[1] The fascinating story of this theorem is given in Robin Wilson, *Four Colours Suffice: How the Map Problem Was Solved*, Penguin 2003.

This is also odd. Hence $n^2 + 3n + 7$ is odd for all integers n.

Remark 22.2

As you can see, this **method of cases** involves exhausting all the possibilities and so this method is also know as **exhaustion**.

Example 22.3

Show that $n^3 - n$ is a multiple of 3 for all $n \in \mathbb{N}$. (We shall see another method to show this in Chapter 24, Induction.)

The number n can be written in the form $n = 3k + r$ for some $k \in \mathbb{N}$, and $r = 0$, 1 or 2. (This is obvious, but will be proved rigorously in Chapter 28.) What we shall do is show that for the three cases, $r = 0$, $r = 1$ and $r = 2$, the statement is true.

Case 1: $r = 0$. In this case, $n = 3k$. Then

$$n^3 - n = (3k)^3 - (3k) = 27k^3 - 3k = 3(9k^3 - k).$$

Hence, n is a multiple of 3.

Case 2: $r = 1$. Here $n = 3k + 1$ so

$$\begin{aligned} n^3 - n &= (27k^3 + 27k^2 + 9k + 1) - (3k + 1) \\ &= 27k^3 + 27k^2 + 6k = 3(9k^3 + 9k^2 + 2k). \end{aligned}$$

So, n is a multiple of 3.

Case 3: $r = 2$. Here $n = 3k + 2$ so

$$\begin{aligned} n^3 - n &= (27k^3 + 54k^2 + 36k + 8) - (3k + 2) \\ &= 27k^3 + 54k^2 + 33k + 6 = 3(9k^3 + 18k^2 + 11k + 2). \end{aligned}$$

Again, n is a multiple of 3.

Exercise 22.4

Streamline the proof by calculating the general form $n^3 - n = (3k + r)^3 - (3k + r)$ first and then using $r = 0$, 1, 2.

The modulus function

Cases can also be used in definitions. An important function in mathematics is the modulus function. This can be defined using cases.

Definition 22.5

*The **modulus** of a real number x, denoted $|x|$, is defined by*

$$|x| = \begin{cases} x, & \text{for } x \geq 0, \\ -x, & \text{for } x < 0. \end{cases}$$

Here we have two cases, one for when x is greater than or equal to 0 and the other when x is less than 0. The point is that we are removing a negative sign if there is one. The definition can also be given as $|x| = \sqrt{x^2}$. (Don't forget that the square root of a number is always positive.)

Examples 22.6

(i) $|3| = 3$,

(ii) $|-3| = 3$,

(iii) $|-5/2| = 5/2$,

(iv) $|\pi| = \pi$,

(v) $|0| = 0$.

An extremely useful theorem in many areas of mathematics is called the Triangle Inequality.

Theorem 22.7 (Triangle Inequality)

Suppose that $x, y \in \mathbb{R}$. Then

$$|x + y| \le |x| + |y|.$$

Proof. As the modulus function is defined using cases and is dependent on the sign of the variable, then it should come as no surprise that the cases in our proof depend on whether x and y are both positive, negative, or differ in sign.

Case 1: Suppose that $x \ge 0$ and $y \ge 0$. Then, $x + y \ge 0$, so

$$|x + y| = x + y = |x| + |y|.$$

Thus, in this case, the statement holds.

Case 2: Suppose that $x < 0$ and $y < 0$ (and so $-x = |x|$ and $-y = |y|$. Check the text!). Then, $x + y < 0$, so

$$|x + y| = -(x + y) = -x + (-y) = |x| + |y|.$$

Thus, the statement holds.

Case 3: Suppose that one of x and y is positive and the other negative. Without loss of generality we can assume that $x \ge 0$ and $y < 0$. This gives us two subcases to consider: $x + y \ge 0$ and $x + y < 0$.

Subcase (a): Suppose that $x + y \ge 0$. Then

$$\begin{aligned}
|x + y| &= x + y \\
&\le x + (-y) \text{ because } y < 0 < -y \\
&= |x| + |y|.
\end{aligned}$$

Subcase (b): Suppose that $x + y < 0$. Then, since $-x \le x$ for $x \ge 0$, we have

$$|x + y| = -(x + y) = (-x) + (-y) \le x + (-y) = |x| + |y|.$$

$\square$

The reason for calling this the Triangle Inequality becomes clearer when it is generalized to the case of complex numbers.

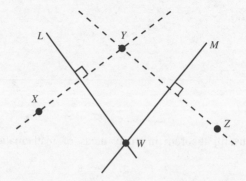

Figure 22.1 Figure for proof about three points lying on a line or a circle

The importance of cases in extreme examples

The method of cases is important when we consider extreme examples. These will cause us to deal with extreme cases. For example, in geometry if we take three random points in the plane, then they do not necessarily form the vertices of a triangle. They could all lie on a line. We have to watch for these extreme examples.

The following theorem (which is quite a nice theorem in its own right) and its proof will exemplify this.

Theorem 22.8

Let X, Y and Z be points in the plane. Then, all lie on a line or all lie on a circle.

Proof. Denote by L the perpendicular bisector (see page 134) between X and Y, and by M the perpendicular bisector between Y and Z. If the points all lie on a line, then they satisfy the conclusion of the theorem and we are done. If the points are not on a line, then L and M are not parallel and so must meet at a point W, say. See Figure 22.1.

As W is on the perpendicular bisector L it is equidistant from X and Y (i.e. the distance from W to X is the same as W to Y). Similarly it is equidistant from Y and Z. As the distance from W to X and the distance from Y and Z are the same we deduce that the three points lie on a circle centred at W of radius equal to the distance from W to X. □

Exercise 22.9

Apply the methods of Chapters 16 and 18 to this theorem and proof.

As you can see in the proof, an extreme case to be dealt with was that the points could all be on a line. However, the proof fails when one considers another extreme case. What happens if the three chosen points are all the same point – note that the theorem did not say the points had to be distinct – or if two of the points are the same?

If X and Y are the same point, then it is impossible to define the perpendicular bisector. Thus we could not have defined L. Is this is a problem? Is the theorem wrong?

Fortunately, the error in the proof is non-fatal and the statement of the theorem does not need to change. After the first line of the proof about labelling the points X, Y and Z we should add the following.

'If the points are not distinct, then two or more of them coincide. If all three coincide, then the points lie on an infinite number of lines as any line through the point will suffice. If two coincide, then take the unique line through the coincident points and the third points. In any case the conclusion of the theorem is true and we may now assume that the points are distinct.'

In fact we could say the following: 'If the points are not distinct, then at least two are the same point. Hence, taking a line through this and the third (possibly non-distinct) point shows that the points lie on a line. Now for the case where the points are distinct ...'

Exercises

Exercises 22.10

(i)   Prove that the square of any integer is of the form $3k$ or $3k + 1$ for some $k \in \mathbb{Z}$.

(ii)   Prove that the cube of any integer is of the form $9k$, $9k + 1$, or $9k + 8$ for some $k \in \mathbb{Z}$.

(iii)   Prove the following about the modulus function for $x, y \in \mathbb{R}$.

(a) $-x \leq |x|$ and $x \leq |x|$.

(b) $|x| \leq y \iff (-x \leq y$ and $x \leq y) \iff -y \leq x \leq y$.

(c) $|xy| = |x||y|$.

(d) $\left| \dfrac{x}{y} \right| = \dfrac{|x|}{|y|}$ for $y \neq 0$.

(e) $||x| - |y|| \leq |x - y|$.

(f) $|x| = \sqrt{x^2}$.

(iv)   Show that $A \cup B = (A \cap B) \cup (A \backslash B) \cup (B \backslash A)$.

(v)   Suppose $a = bc$ for three real numbers a, b and c. Prove that if two of a, b and c are non-zero, then so is the third.

(vi)   Use Pythagoras' Theorem to prove that for all $0 < \theta < \pi/2$ we have

$$\cos^2 \theta + \sin^2 \theta = 1.$$

(In some sense, this is the only equation relating $\sin \theta$ and $\cos \theta$ so is very valuable.)

Use cases to prove this for *all* $\theta \in \mathbb{R}$. (You will need to define sin and cos for angles greater then $\pi/2$.)

(vii)   Define the **maximum function** $\max(x, y)$ to be given by $\dfrac{x + y + |x - y|}{2}$. Show that this does give the maximum of the two numbers x and y.

(viii) Create a definition for the **minimum function** and show that it does give the minimum of two numbers.

(ix)   Here's another classic brain teaser. It is great because it looks as though we do not have enough information to answer it, and worse still, how can a child's hair colour help in a calculation problem?

Two mathematician friends meet on the street. The first says 'How old are your three sons? Remind me, I've forgotten their ages.' The second says 'Well, the product of their ages is 36 and the sum of their ages is equal to the number of windows in the house over there.' The first thinks for a moment and says 'I can't figure it out.' The second replies 'Oh, I forgot to tell you that my oldest son has blonde hair.' The first replies 'In that case, I know the ages of your sons.'

Can you work out the ages?

Summary

▶ It is sometimes convenient to break a problem into cases.

▶ We may need to analyse separately extreme cases in a problem.

▶ The definition of the modulus function is $|x| = \begin{cases} x, & \text{for } x \geq 0, \\ -x, & \text{for } x < 0. \end{cases}$

▶ $|x| = \sqrt{x^2}$.

▶ Triangle Inequality: For all $x, y \in \mathbb{R}$ we have $|x + y| \leq |x| + |y|$.

CHAPTER 23

Techniques of proof III: Contradiction

Let me never fall into the vulgar mistake of dreaming that
I am persecuted whenever I am contradicted.
Ralph Waldo Emerson, Journal entry, 8 November 1838

The law of the excluded middle asserts that a statement is true or it is false, it cannot be anything in between. We can use this as another method of proof. We assume that the statement is false and proceed logically to show that this gives a statement that we definitely know is false such as $1 = 0$ or the Moon is made of cheese. Thus our assumption must be wrong, the statement *can't* be false – it leads to something ridiculous – so the statement is true.

This method is called proof by contradiction. The name comes from the fact that assuming that the statement is false is later contradicted by some other fact. It is also known by the name **reductio ad absurdum** which when translated means reduction to the absurd.

Simple examples of proof by contradiction

The first example is just to show you the idea of proof by contradiction. The statement is easier to prove by a direct method as we have seen in Theorem 20.1.

Example 23.1

Suppose that n is an odd integer. Then n^2 is an odd integer.

Proof. Assume the contrary. That is, we suppose that n is an odd integer but that the conclusion is false, i.e. n^2 is an even integer.

As n is odd, $n = 2k + 1$ for some $k \in \mathbb{Z}$. Thus $n^2 = (2k + 1)^2 = 4k + 2k + 1$ which contradicts that n^2 is even. Thus our assumption that n^2 is even must be wrong, i.e. n^2 must be odd. $\qquad\square$

The statement above has the form $A \implies B$. In general, if we assume such a statement is false, then we are assuming that 'A and not(B)' as this is the negation of $A \implies B$ (see page 66). To use contradiction we then have to show that 'A and not(B)' leads to something false.

The second example solves a harder problem.

Example 23.2

There are no positive integers x and y such that $x^2 - y^2 = 1$.

Proof. We assume the contrary, i.e. we assume that positive integers exist such that $x^2 - y^2 = 1$. Thus we have $(x + y)(x - y) = 1$. Since x and y are integers, then $x + y$ and $x - y$ are integers too, so we have two cases:

Case 1: $x + y = 1$ and $x - y = 1$. Solving this pair of equations gives $x = 1$ and $y = 0$. This contradicts that x and y are positive. This leaves the second case.

Case 2: $x + y = -1$ and $x - y = -1$. Solving these we find that $x = -1$ and $y = 0$, again contradicting that the two integers are positive. □

As you can see, the statement was '$\nexists x, y \in \mathbb{N}(x^2 - y^2 = 1)$'. We assumed that the result was false, in other words, that its negation '$\exists x, y \in \mathbb{N}(x^2 - y^2 = 1)$' is true.

Example 23.3

The sum of a rational and an irrational number is an irrational number.

We can see this better if we explicitly write the statement as an implication. In other words, 'If x is rational and y is irrational, then $x + y$ is irrational.' Using the fact that '$A \Longrightarrow B$' has negation 'A and $\text{not}(B)$' we assume that x is rational, y is irrational, and $x + y$ is not irrational are all true.

Proof. Assume to the contrary, that is x is rational, y is irrational and $x + y$ is rational. Since x is rational $x = p/q$ for some integers p and q. Similarly, $x + y$ rational implies that $x + y = r/s$ for some integers r and s.

We see

$$x + y = \frac{r}{s}$$
$$\Longrightarrow \quad \frac{p}{q} + y = \frac{r}{s}$$
$$\Longrightarrow \quad y = \frac{r}{s} - \frac{p}{q}$$
$$\Longrightarrow \quad y = \frac{rq - ps}{sq}.$$

But $\dfrac{rq - ps}{sq} \in \mathbb{Q}$ which contradicts that y is irrational. Hence the statement is true. □

Exercise 23.4

Consider the statement 'The product of a rational and an irrational number is an irrational number.' Prove this statement or give a counterexample. If you give any counterexamples, can you change the statement slightly so that you do have a true statement?

Example 23.5

The equation $x^7 + 3x^3 + 5$ has no rational roots.

Assume to the contrary that x is a root and is rational. Thus $x = p/q$ where p and q are integers and this quotient is in its simplest terms. (That is, we can't divide top and bottom by the same number greater than 1.)

Then we have

$$\left(\frac{p}{q}\right)^7 + 3\left(\frac{p}{q}\right)^3 + 5 = 0$$
$$p^7 + 3p^3q^4 + 5q^7 = 0.$$

We can consider what happens when p and q are odd and even. There are four cases to consider.

Case 1: If p and q are both even, then p/q is not in its simplest form. Thus, we get a contradiction.

Case 2: If p and q are both odd, then the left-hand side of $p^7 + 3p^3q^4 + 5q^7 = 0$ is odd, while the right-hand side is even. (Check this with an exercise from Chapter 20.) This is a contradiction.

Case 3: If p is even and q is odd, then the left-hand side of the above equation is odd, while the right is even. Again, this is a contradiction.

Case 4: If p is odd and q is even, then, again, the left-hand side is odd and the right-hand side is even. A contradiction.

Thus, x is not rational.

The irrationality of the square root of 2

Now we use proof by contradiction to show a classic theorem and proof of mathematics: $\sqrt{2}$ is an irrational number. That is, it cannot be written as the quotient of two integers.

The negation of this statement is that $\sqrt{2}$ is rational. We proceed to show that this assumption leads to an impossible statement.

Theorem 23.6

The square root of 2 is irrational, i.e. cannot be written in the form m/n where m and n are integers.

Proof. Suppose to the contrary that $\sqrt{2} = m/n$ where m and n are integers. Without loss of generality we can assume that this quotient is in its simplest terms. Then we have,

$$\sqrt{2} = \frac{m}{n}$$
$$2 = \left(\frac{m}{n}\right)^2, \quad \text{by squaring both sides,}$$
$$2 = \frac{m^2}{n^2}$$
$$2n^2 = m^2.$$

This implies that m^2 is even since it is the product of 2 and n^2. We now have two choices for m: it can be even or it can be odd. If m is odd, then m^2 is also odd, by Theorem 23.1. (Did you look back to check that the referenced theorem gives this?) Hence we must have m even.

So, $m = 2k$ for some integer k. Then using the equation $2n^2 = m^2$ we get

$$2n^2 = (2k)^2$$
$$= 2^2 k^2$$
$$= 4k^2$$
$$n^2 = 2k^2.$$

By reasoning similar to that above we conclude that n has to be even as well, i.e. $n = 2j$ for some integer j. However, we assumed that the quotient m/n was in its lowest form; this has been shown to be not the case:

$$\sqrt{2} = \frac{m}{n} = \frac{2k}{2j} = \frac{k}{j}.$$

Thus we conclude that $\sqrt{2}$ cannot be written as a quotient of integers.   □

Exercises 23.7

(i)   Show that $\sqrt{3}$ is irrational.

(ii) Show that $\sqrt{5}$ is irrational.

(iii) Apply the proof to the *non-example* '$\sqrt{4}$ is irrational'. What happens? Where does the proof go 'wrong' in this case.

(iv) Can you generalize the method to $\sqrt{p}$ is irrational where p is a prime?

How to spot a proof by contradiction

We do not spot proofs by contradiction as such but automatically turn to the method if we cannot prove the statement directly.

For example, to prove that something does not exist, we assume that it does and aim for a contradiction. And vice versa.

The point is that it is difficult to do operations with something that does not exist. Assuming something exists means we can apply operations. For example, to show something is irrational, it is easier to assume it is rational because then we can write it in the form p/q for integers p and q.

How to write a proof by contradiction

(i)   State that you are assuming the statement is false. Seasoned mathematicians will recognize that the proof will be by contradiction.

(ii) Write out what the statement being false means using negation.

(iii) Work out what this would imply until you find a contradiction.

(iv) Announce that a contradiction has been found.

Exercises

Exercises 23.8

(i)   Show that the solutions of the equation $x^5 - 2x^3 - 3 = 0$ are all less than 2. (Hint: It is easier to find roots less than 0 so change to a different variable.)

(ii) Prove that for all integers x and y if xy is odd, then x and y are both odd.

(iii) Prove by contradiction that there exists an infinite number of rational numbers between 0 and 1. (Hint: Consider Theorem 20.4.)

(iv) Show that proof by contradiction for P is, logically speaking,

$$P \text{ is equivalent to (not } P \Longrightarrow (Q \text{ and not } Q)).$$

(v)   Prove that there are no positive integer solutions to $x^2 + x + 1 = y^2$.

(vi) Prove that there is no greatest rational number less than $\sqrt{2}$.

(vii) For all rational numbers x and y with $x < y$ prove that there exists an irrational number z such that $x < z < y$.

(viii) Prove the above but with x, y irrational and z rational.

(ix) Give an example of a sum of two irrational numbers that is rational. Is $\sqrt{2} + \sqrt{3}$ rational or irrational? Explain.

(x)   Show that $\log_2 3$ is irrational. (Hint: By definition, $\log_2 3$ is the number x such that $2^x = 3$.)

(xi) Prove or disprove: If x is irrational, then $\sqrt{x}$ is irrational.

(xii) Is $\sqrt[3]{2}$ irrational or rational? Either way, can you generalize your statement?

(xiii) Suppose that x and y are positive integers. Show that $\sqrt{x^2 + y^2} \neq x + y$.

Summary

▶ In proof by contradiction we assume that the negation of the statement is true and from that deduce something that is obviously false.

▶ The square root of 2 is irrational.

▶ Write statements such as 'The sum of a rational and an irrational number is an irrational number' as implications.

CHAPTER 24

Techniques of proof IV: Induction

One thing leads to another.
Anon.

Induction is a very powerful technique used regularly by mathematicians.[1] Initially, it can be confusing because it *looks* like we assume what is to be proved. As we know, that never proves theorems. On the plus side, spotting when to use it is easy and we need only check two conditions to apply it.

Induction is applied when we have an infinite number of statements indexed by the natural numbers such as

$n^5 - n$ is even for all $n \in \mathbb{N}$'.

It is not sufficient to prove this for a sample of natural numbers, whether that sample involves hundreds, millions or even billions of numbers; we have to prove it for *all n*.

With induction we don't prove the statements directly. What we do is perhaps best described by analogy with domino toppling. As you are probably aware, this is where dominoes are standing on their ends in such a way that when you push the first one over, it knocks the second domino over, that in turn knocks the third down, and so on. Provided the dominoes are arranged so that each knocks down the next, then all of them will fall.

The process of induction is that we prove that 'if the kth statement is true, then the $k+1$th statement is true', i.e. the truth of one statement implies the truth of the next one. This is analogous to one domino knocking down the next one. So, if the first statement is true (push the first domino), then all the statements are true (all the dominoes get knocked down).

The Principle of Mathematical Induction

Let's begin to make the idea precise. First, the Principle of Mathematical Induction requires that we have a sequence of statements indexed by the natural numbers. There are plenty of these, as the following examples show. Note that they are from diverse areas of mathematics: divisibility, summation, inequalities. These statements will all be proved

---

[1] There is a principle of induction in experimental science which is different to the one we are about to discuss.

later using induction. Hopefully, by showing you this scope you will get an impression of the power of induction – it really is an important tool in the mathematician's toolkit.

Examples 24.1

(i)   The expression $6^n - 1$ is divisible by 5 for all $n \in \mathbb{N}$. (We say a is **divisible** by b if there is no remainder when we divide a by b. It means that $a = bq$ for some $q \in \mathbb{Z}$. We shall see more of this in Chapter 27.)

(ii) We have $\sum_{i=1}^{n} i = \frac{1}{2}n(n + 1)$ for all $n \in \mathbb{N}$.

(iii) The inequality $2^{n-1} \leq n!$ is true for all $n \in \mathbb{N}$.

After reading these did you behave like a mathematician and try out a few cases to check that the statements were at least plausible?

Let's check some cases. For (i) we have

$$n = 1: \quad 6^1 - 1 = 6 - 1 = 5. \qquad\qquad \text{Divisible by 5.}$$
$$n = 2: \quad 6^2 - 1 = 36 - 1 = 35. \qquad\quad\;\, \text{Divisible by 5.}$$
$$n = 3: \quad 6^3 - 1 = 216 - 1 = 215. \qquad\;\, \text{Divisible by 5.}$$
$$n = 10: \quad 6^{10} - 1 = 60466176 - 1 = 60466175. \quad \text{Divisible by 5.}$$

Notice that I did the first few cases and then threw in a higher case. This would seem to indicate that the statement is reasonable – we haven't found a counterexample yet!

Exercise 24.2

Check the other two statements above for small n to help convince yourself that they are true.

This type of checking may seem pointless – 'Aren't we going to prove the statements are true in a minute?' Yes, however, what it does is engage the brain in the statement and ensures that you see what is there – not what you think is there.

We will use the notation $A(n)$ to denote the statement for a particular n. In (i) we have, for example, $A(3)$ is just 'The expression $6^3 - 1$ is divisible by 5.'

Now, ladies and gentlemen, I give you the grand theorem!

Theorem 24.3 (Principle of Mathematical Induction)

Let $A(n)$ be an infinite collection of statements with $n \in \mathbb{N}$. Suppose that

(i)   *$A(1)$ is true, and*

(ii) *$A(k) \Longrightarrow A(k + 1)$, for all $k \in \mathbb{N}$.*

Then, $A(n)$ is true for all $n \in \mathbb{N}$.

Proof. We aim for a contradiction. Assume the conclusion is false and let j be the smallest natural number such that $A(j)$ is false. By assumption (i) we have that $j > 1$. Now we note that $A(j - 1)$ has to be true as j is the smallest possible. Hence by assumption (ii) we have that $A(j)$ is true. This is a contradiction. $\qquad\qquad\square$

We need some definitions. The reasons for making them will become clearer later.

Figure 24.1 Toppling dominoes

- Checking condition (i) is called the **initial step**.
- Checking condition (ii) is called the **inductive step**.
- Assuming that $A(k)$ is true for some k in (ii) is called the **inductive hypothesis**.

The idea is represented in Figure 24.1 where each statement is seen as a domino.

Examples of induction

Let's see the principle in action. The next example is a real classic of mathematics: 'What is the sum of the first n numbers?'

In a famous story, when the great mathematician Johann Carl Friedrich Gauss (1777–1855) was at school, his teacher asked the class to sum all the numbers up to 100 – presumably in the hope that it would keep them occupied for a lesson. While his classmates worked through the exercise in the laborious, obvious way, Gauss thought like a mathematician and avoided the drudgery by producing a simple method for calculating the sum of the first 100 numbers, thwarting his teacher's ambition for a quiet, easy lesson for himself.

Gauss' method can be applied to produce a formula for all numbers, not just 100. This formula is that in Examples 24.1(ii) above. We shall prove the truth of this formula by induction rather than use his method of proof.[2]

Example 24.4

We shall prove that $\sum_{i=1}^{n} i = \frac{1}{2}n(n + 1)$ for all $n \in \mathbb{N}$.

This statement is indexed by n, so we can call $A(n)$ the statement

$$\sum_{i=1}^{n} i = \frac{1}{2}n(n + 1).$$

Our aim is to show that $A(n)$ is true for each n.

Let us check condition (i), the initial case. This corresponds to pushing the first domino. This is easy:

When $n = 1$, then

$$\sum_{i=1}^{n} i = \sum_{i=1}^{1} i = 1$$

[2] Gauss' method was to note that we can pair up the numbers so that the sums of pairs is constant, i.e. $1 + 100$, $2 + 99$, $3 + 98$, and so on. The rest of the formula is easy: we have 50 pairs (i.e. $100/2$) summing to 101. The formula in the general case uses the same argument.

and

$$\tfrac{1}{2}n(n+1) = \tfrac{1}{2} \times 1 \times (1+1) = 1.$$

Now for the inductive step. We now want to show that if the statement $A(k)$ is true, then $A(k+1)$ is true as well. This is the same as domino number k knocking over domino $k+1$.

Let us assume that $A(k)$ is true for some arbitrary k. (This is a single k; the point is that it is arbitrary. We make no assumptions for any other number n, just this particular k.) That is,

$$\sum_{i=1}^{k} i = \tfrac{1}{2}k(k+1).$$

We investigate what implications this has for $A(k+1)$. First we write down the form of $A(k+1)$:

$$\sum_{i=1}^{k+1} i = \tfrac{1}{2}(k+1)((k+1)+1).$$

How would a mathematician prove such an equality? Experience has taught us that we should pick the complicated side of an expression and reduce it to get the other side. For a proof by induction of a statement involving a sum the complicated side is usually the one with the summation (as it probably has more terms).

We have

$$\sum_{i=1}^{k+1} i = \left(\sum_{i=1}^{k} i \right) + (k+1), \text{ by the definition of summation,}$$

$$= \left(\tfrac{1}{2}k(k+1) \right) + (k+1), \text{ by the inductive hypothesis,}$$

$$\text{i.e. as } A(k) \text{ is true,}$$

$$= \left(\tfrac{1}{2}k+1 \right)(k+1)$$

$$= \tfrac{1}{2}(k+2)(k+1)$$

$$= \tfrac{1}{2}(k+1)((k+1)+1).$$

Thus, we have shown that if

$$\sum_{i=1}^{k} i = \tfrac{1}{2}k(k+1)$$

is true, i.e. $A(k)$ is true, then

$$\sum_{i=1}^{k+1} i = \tfrac{1}{2}(k+1)((k+1)+1)$$

is true, i.e. $A(k+1)$ is true. In other words, we have shown that all the dominoes are lined up so that if the kth falls, then the $(k+1)$th falls.

Thus, by the Principle of Mathematical Induction the statement is true for all $n \in \mathbb{N}$.

Remarks 24.5

(i) Note the structure. We assume that $A(k)$ is true for an arbitrary k. Then we look at the $A(k + 1)$ case, and tease it apart so that $A(k)$ can be used. We *don't* assume that the $A(k + 1)$ case is also true – this is a common error by beginners – we show that it is true *when* $A(k)$ is true.

(ii) Another problem for novice mathematicians is that induction *seems* to violate the principle that we do not assume what we are trying to prove. I felt this myself as a student.

However, it is vital to grasp the subtlety in induction if it is to be applied confidently. It should be noted that the statement we wish to prove is about something holding for *all n*. In our assumption in condition (ii) we assume that the statement holds for *one* particular n which we call k. Ok, that one particular n is arbitrary, it is absolutely any n you like, but it is still just one, by itself, no other assumptions made.

Example 24.6

The expression $6^n - 1$ is divisible by 5 for all $n \in \mathbb{N}$.

We shall do this with less explanation than the previous example – this example will be more like a model solution in a book or submitted for an assignment.

Initial step: The statement is true for $n = 1$ as $6^n - 1 = 6^1 - 1 = 5$.

Inductive step: Assume the statement is true for some[3] $k \in \mathbb{N}$; this means that $6^k - 1 = 5m$ for some $m \in \mathbb{N}$. Then

$$6^{k+1} - 1 = 6(6^k) - 1$$
$$= 6(5m + 1) - 1, \text{ by the inductive hypothesis,}$$
$$= 30m + 6 - 1$$
$$= 5(6m + 1).$$

This is divisible by 5 and so the statement is true for $k + 1$. Hence, by the Principle of Mathematical Induction, the statement is true for all $n \in \mathbb{N}$.

Now for the third example.

Example 24.7

We show that $2^{n-1} \leq n!$ for all $n \in \mathbb{N}$.

Initial step: For $n = 1$ we have $2^{n-1} = 2^0 = 1$ and $n! = 1! = 1$. Hence, $2^{n-1} \leq n!$ for $n = 1$.

Inductive step: Assume the statement is true for some $k \in \mathbb{N}$, that is $2^{k-1} \leq k!$.
Then, for $n = k + 1$:

$$2^{(k+1)-1} = 2^k$$
$$= 2(2^{k-1})$$
$$\leq 2(k!), \text{ by the inductive hypothesis,}$$
$$\leq (k + 1)k!, \text{ as } 2 \leq k + 1,$$
$$= (k + 1)!$$

---

[3] Note: Remember it is *some* k – an arbitrary single k – not *all* k.

This shows that the statement is true for $n = k+1$. Hence, by the Principle of Mathematical Induction, the statement is true for all $n \in \mathbb{N}$.

Note that we took one side of the equality in the statement for $n = k+1$ and played with that. You could just as easily start with the other side, just as long as you don't assume $A(k + 1)$.

Let's try something which is more of a problem and which allows us to exercise our brains with some real mathematical thinking.

Example 24.8

Find a formula for the sum of the first n odd numbers.

We have two problems wrapped up in one. First we must find the formula by some method and then prove it holds for all n (which, it should come as no surprise, will be by induction).

For the first part we can refer to our problem-solving methods (see Chapter 5). A good method of attack is to rewrite the problem in symbols. In symbols the sum of the first n odd numbers is $\sum_{i=1}^{n} (2i - 1)$. Let's be good mathematicians: try the first few cases.

These results are tabulated below.

n	1	2	3	4	5
$\sum_{i=1}^{n}(2i - 1)$	1	4	9	16	25

It doesn't take much to see what the pattern is: The sum of the first n odd numbers is n^2. Well, that's what it looks like – it could be that the next one doesn't fit the pattern. Nonetheless, we have a good conjecture to pursue:

$$\sum_{i=1}^{n}(2i - 1) = n^2 \text{ for all } n \in \mathbb{N}.$$

This is a nice, simple formula which we have guessed from investigating the first few cases. This is truly thinking like a mathematician. Of course, we need to show that it is true!

The initial step has been done in our tabulated calculations. Let's do the inductive step. Assume the statement is true for some k, i.e. $\sum_{i=1}^{k}(2i - 1) = k^2$. Then,

$$\sum_{i=1}^{k+1}(2i - 1) = \left(\sum_{i=1}^{k}(2i - 1)\right) + (2(k + 1) - 1)$$
$$= k^2 + (2k + 2 - 1), \text{ by the inductive hypothesis,}$$
$$= k^2 + 2k + 1$$
$$= (k + 1)^2.$$

Thus, if the kth statement is true, then so is the $(k + 1)$th statement. Hence, our formula is true by the Principle of Mathematical Induction.

Notice that once again in the summation for the $k + 1$ case all we do is split off the last summand and follow by a simple application of the k case.

Remark 24.9

In writing the above a mathematician would remove all mention of the calculations of the initial few cases that led to the formula. Instead he or she would state the formula and proceed to show it was true. This is known as **covering your tracks** because all information on how we got there has been removed.

How to spot an induction proof

Obviously, to spot that induction could be used you must be suspicious of any statement indexed by the set of natural numbers. For example, 'For every natural number ...', or '... for all $n \in \mathbb{N}$'.

Sometimes the statements can be disguised. For example, consider the exercise

'Show that $n^2 - 1$ is divisible by 8 when n is an odd natural number.'

At first glance this does not appear to be indexed by the natural numbers as none of the even numbers are used. However, the statement *is* indexed by the naturals as the set of odd numbers can be matched up with the naturals. That is, 1 is the *first* odd number, 3 is the *second*, 5 is the *third*, and so on. We could rewrite the statement as

'Show that $(2n - 1)^2 - 1$ is divisible by 8 when n is a natural number.'

How to write an induction proof

The method for writing a proof by induction is simple:

(i)   Announce that you are using induction.
(ii) Do the initial case.
(iii) State that you are assuming that the statement is true for some k. Writing out the statement to use it later is often helpful.
(iv) Use the truth of the statement for k in the proof of the statement for $k + 1$. Often this will mean breaking a mathematical expression into two pieces one of which involves the case for k. Be sure to indicate at which point you use the inductive hypothesis.
(v)   State the conclusion: 'By the Principle of Mathematical Induction the statement is true.' That way, the reader knows the proof is over.

Exercises

Exercises 24.10

(i)   Prove by induction that $\sum_{i=1}^{n} i^2 = \frac{n}{6}(n + 1)(2n + 1)$.
(ii)   Show that $2n \leq 2^n$ for all natural numbers n.
(iii) Prove that $3^{2n} - 1$ is divisible by 8 for all natural numbers n.
(iv)   Prove that 17 divides $3^{4n} + 4^{3n+2}$ for all $n \in \mathbb{N}$.
(v)   Show that $\sin nx \leq n \sin x$ for all natural numbers n and $0 \leq x \leq \frac{\pi}{2}$.

(vi) Prove the Binomial Theorem, that is,

$$(x + y)^n = \sum_{r=0}^{n} \binom{n}{r} x^{n-r} y^r \text{ for all } n \in \mathbb{N}.$$

$$\left(\text{Recall that } \binom{n}{r} = \frac{n!}{r!(n-r)!} \text{ for all } 0 \leq r \leq n. \right)$$

(vii) Show that $n^2 - 1$ is divisible by 8 when n is an odd natural number.

(viii) Prove Leibniz's Theorem for repeated differentiation of a product: If u and v are functions of x, then prove that

$$\frac{d^n}{dx^n}(uv) = u_0 v_n + \binom{n}{1} u_1 v_{n-1} + \binom{n}{2} u_2 v_{n-2} + \cdots + \binom{n}{r} u_r v_{n-r}$$
$$+ \cdots + u_n v_0,$$

for all $n \in \mathbb{N}$, where u_i and v_j denote $\dfrac{d^i u}{dx^i}$ and $\dfrac{d^j v}{dx^j}$ respectively.

(You will need to use

$$\binom{k}{r-1} + \binom{k}{r} = \binom{k+1}{r}$$

but proving this identity is true should not pose much difficulty.)

(ix) Prove that $\displaystyle\sum_{r=1}^{n} r^3 = \left(\sum_{r=1}^{n} r \right)^2.$

(x) Show that $\binom{n}{r} / n$ is an integer for all $1 \leq r \leq n - 1$.

(xi) Let X be a finite set with n elements. Show that X has 2^n distinct subsets.

(xii) This exercise will, in effect, generalize the result of Exercise (iii).

    (a) Show that $x^n - 1$ is divisible by $x - 1$ for all $n \in \mathbb{N}$ and where $x \neq 1$ is a natural number.

    (b) Find a formula for $\dfrac{x^n - 1}{x - 1}$ and use induction to show it holds for all n.

(xiii) Consider the statement '$A(n)$: $2^n < 2^{n-1}$'. Prove that $A(n)$ is *false* for all $n \in \mathbb{N}$. Show that the inductive step holds, i.e. that $A(k) \Longrightarrow A(k+1)$. Note that this shows that we need the initial case to be true to show the statement holds for all n. This is like 'The Moon is made of cheese implies that the Moon is a tasty snack' from Chapter 7. That is, '$A(n) \Longrightarrow A(n+1)$' can be true even if both $A(n)$ and $A(n+1)$ are false.

(xiv) Prove the statement on negation of quantifiers from Chapter 11:

    To negate a statement of the form $Q_1 x_1 Q_2 x_2 \ldots Q_n x_n P(x_1, x_2, \ldots, x_n)$, where Q_i is $\forall$ or $\exists$ for $1 \leq i \leq n$, we do the following:

    (a) Change every $\forall$ to $\exists$ and every $\exists$ into $\forall$.

    (b) Replace P by its negation.

(xv) Show that $\sum_i^n |x_i| \geq |\sum_i^n x_i|$. (This is the generalized Triangle Inequality.)

(xvi) Suppose $x_1, x_2, \ldots, x_n$ are non-negative real numbers. Prove that

$$\frac{x_1 + x_2 + \cdots + x_n}{n} \geq \sqrt[n]{x_1 x_2 \ldots, x_n}.$$

Summary

- ▶ The principle of mathematical induction is useful for proving statements indexed by the natural numbers.
- ▶ Think of applying induction when you see statements indexed by the natural numbers.
- ▶ Induction is used extensively in mathematics, e.g. summations, inequalities, divisibility, . . .
- ▶ First prove $A(1)$, then show $A(k) \implies A(k + 1)$ for some (arbitrary but fixed) $k \in \mathbb{N}$.
- ▶ Remember, it's *some k*, not *all k*.

CHAPTER 25

More sophisticated induction techniques

Simplicity is the ultimate sophistication.

Leonardo da Vinci

In this chapter we investigate more sophisticated versions of induction. There are three variants we shall be most interested in.

(i)   We use a different initial case. Rather than show that $A(1)$ is true we show, for instance, $A(7)$ or $A(15)$ is true. Thus $A(n)$ is true for all $n \geq 7$ or all $n \geq 15$ respectively.

(ii) We change the inductive step to '$A(k-1)$ *and* $A(k)$ imply $A(k+1)$.' This requires us to have as initial case that $A(1)$ and $A(2)$ are true.

(iii) We change the inductive step to '$A(j)$ true for all $1 \leq j \leq k$ implies $A(k+1)$ true.' We use initial case $A(1)$ true or some other initial case like (i) above.

All three can be referred to as the Principle of Mathematical Induction and, in addition, the latter two are sometimes called the **Principle of Strong Mathematical Induction**.

First variant

We do not need to start with $n = 1$ as the initial case. For example, for statements $A(n)$ the first few cases may be false. If we can show

(i) $A(r)$ is true for some $r \in \mathbb{N}$, and
(ii) $A(k) \Longrightarrow A(k+1)$ for all $k \geq r$,

then the statement is true for all $A(n)$ with $n \geq r$. Observe that our main change to induction is really the initial step – the change of the range of values in the inductive step is minimal.

Example 25.1

Prove that $n^2 \leq 2^{n-1}$ for all $n \in \mathbb{N}$ such that $n \geq 7$.

   This is interesting because that the statement is true for $n = 1$, but it is not true for $n = 2, 3, 4, 5$, or 6.

Initial step: The initial case here is $n = 7$. We have

$$n^2 = 7^2 = 49 < 64 = 2^6 = 2^{n-1}.$$

Thus, the statement is true for $n = 7$.

Inductive step: Assume that the statement is true for some $k \in \mathbb{N}$ with $k \geq 7$, i.e. $k^2 \leq 2^{k-1}$ for $k \geq 7$. We have

$$
\begin{aligned}
(k+1)^2 &= k^2 + 2k + 1 \\
&\leq k^2 + 2k + k, \text{ as } k \geq 7, \\
&= k^2 + 3k \\
&\leq k^2 + k \times k, \text{ as } k \geq 7, \\
&= 2k^2 \\
&\leq 2 \times 2^{k-1}, \text{ by inductive hypothesis}, \\
&= 2^k \\
&= 2^{(k+1)-1}.
\end{aligned}
$$

Therefore, the statement is true for $n = k + 1$. Hence, by the Principle of Mathematical Induction it is true for all $n \geq 7$.

Note that the above list of equalities and inequalities is not obvious, the originator of the proof has covered their tracks.

Exercises 25.2

(i)   Think like a mathematician: where does the proof go wrong for $n < 7$? For which values of k is the argument of the inductive step false?

(ii) Prove that $2^n < n!$ for $n \geq 4$. Show that the cases $n = 1, 2$ and 3 are all false. Just as above, where does the proof go wrong and for which values of k is the inductive step not true?

(iii) Choose n distinct points on a circle and connect them in order to produce a polygon. Show that the interior angles add to $180(n - 2)$ degrees for $n \geq 3$.

Second variant

Another variant is to change the inductive step and use $A(k - 1)$ and $A(k)$ true for $k \geq 2$ to show that $A(k + 1)$ is true. To do this we also need to change the initial case to $A(1)$ and $A(2)$ true. (Do you see why?) Then $A(n)$ is true for all $n \in \mathbb{N}$.

Example 25.3

Suppose that $x_1 = 3$ and $x_2 = 5$ and that for $n \geq 3$

$$x_n = 3x_{n-1} - 2x_{n-2}.$$

Show that $x_n = 2^n + 1$ for all $n \in \mathbb{N}$.

Initial step: $A(1)$ and $A(2)$ are true:

$$n = 1: \ 2^n + 1 = 2^1 + 1 = 3 = x_1,$$
$$n = 2: \ 2^n + 1 = 2^2 + 1 = 5 = x_2.$$

Inductive step: Suppose that the statement is true for x_{k-1} and x_k for some $k \geq 2$. That is,

$$x_{k-1} = 2^{k-1} + 1 \text{ and}$$
$$x_k = 2^k + 1.$$

Now consider the case of $n = k + 1$. We have

$$\begin{aligned}
x_{k+1} &= 3x_{(k+1)-1} + 2x_{(k+1)-2}, \text{ by definition,} \\
&= 3x_k - 2x_{k-1} \\
&= 3(2^k + 1) - 2(2^{k-1} + 1), \text{ by the inductive hypothesis,} \\
&= 3 \times 2^k - 2^k + 1 \\
&= 2 \times 2^k + 1 \\
&= 2^{k+1} + 1.
\end{aligned}$$

Hence, by the Principle of Strong Mathematical Induction the statement is true for all $n \in \mathbb{N}$.

The definition of x_n is known as an **inductive definition**. That is, x_n will depend on x_i where $i < n$.

Exercises 25.4

(i) Let $x_1, x_2, x_3, \ldots$ be a sequence of numbers where $x_1 = 1$, $x_2 = 3$ and $x_n = x_{n-1} + x_{n-2}$ for $n \geq 3$. Show that $x_n < \left(\frac{7}{4}\right)^n$ for all natural numbers n.
(ii) Why do we need $A(2)$ true in addition to $A(1)$ in this variant of induction?

Third variant

The third variant combines variant (i) and a generalization of the inductive step in variant (ii). We need to show that

(i) $A(r)$ is true for some $r \in \mathbb{N}$,
(ii) for all $k \geq r$, $A(j)$ true for all $r \leq j \leq k$ implies that $A(k + 1)$ is true.

Then we can deduce that $A(n)$ is true for all $n \in \mathbb{N}$ with $n \geq r$.

So, for example, if $r = 7$, then in the inductive step we use $A(7), A(8), A(9), \ldots, A(k)$ are true to imply that $A(k + 1)$ is true.

We can use this to prove the following classic theorem.

Theorem 25.5 (Fundamental Theorem of Arithmetic)

Every natural number greater than 1 is a product of primes. That is,

$$n = p_1^{a_1} p_2^{a_2} p_3^{a_3} \cdots p_r^{a_r},$$

for distinct primes $p_1, \ldots, p_r$ and exponents $a_1, \ldots, a_r \in \mathbb{N}$.

Proof. I'll assume that you applied the techniques of earlier chapters to get to grips with the theorem and provide examples and convince yourself it is true. On with the proof . . .

Let $A(n)$ be the statement 'n is either prime or a product of primes'.

Initial step: $A(2)$ is true since 2 is a prime number.

Inductive step: Suppose the statements $A(2)$, $A(3)$, . . . , $A(k)$ are all true for some k. We show that $A(k+1)$ is true. Now, obviously, $k+1$ is either a prime or it is not. If it is, then $A(k+1)$ is true. If $k+1$ is not prime, then $k+1 = xy$ where x and y are natural numbers greater than 1 and less than $k+1$. Therefore, by the inductive hypothesis $A(x)$ and $A(y)$ are true. That is, x and y are products of primes: $x = p_1^{a_1} p_2^{a_2} p_3^{a_3} \ldots p_r^{a_r}$ and $y = q_1^{b_1} q_2^{b_2} q_3^{b_3} \ldots q_s^{b_s}$.

Hence, $k+1 = xy$ is a product of primes. They can be assumed to be distinct since if $p_i = q_j$ for some i and j, then we can take $a_i + b_j$ as the exponent of p_i and of course discard q_j.

Hence we have $A(k+1)$ is true. By the Principle of Mathematical Induction the statement is true for all $n \in \mathbb{N}$. $\qquad\square$

Exercises 25.6

(i)   Use the techniques from previous chapters to understand the proof. Check every assertion and use extreme cases such as 0 and 1 to test the statement and proof. (Note that the statement does not apply to 0 and 1, the point is to gain deeper insight and see *why* the statement is not for true for these cases.)

(ii) Prove that this factorization into primes is unique. That is, if $n = p_1^{a_1} p_2^{a_2} \ldots p_r^{a_r}$ and $n = q_1^{b_1} q_2^{b_2} \ldots q_s^{b_s}$ where the p_i are distinct and $a_i \geq 1$, and the q_j are distinct and $b_j \geq 1$, then $r = s$ and for each i there exists unique j such that $q_j = p_i$.

Exercises

Exercises 25.7

(i)   Let $x_1 = 1$, $x_2 = 1$ and $x_n = x_{n-1} + x_{n-2}$ for $n \geq 3$. This sequence of numbers are known as the **Fibonacci numbers**. Surprisingly, they occur quite regularly in nature.

(a) Write out $x_1, x_2, x_3, x_4, x_5, x_6$ and x_7.

(b) Show that
$$x_n = \frac{1}{\sqrt{5}} \left(\frac{1+\sqrt{5}}{2} \right)^n - \frac{1}{\sqrt{5}} \left(\frac{1-\sqrt{5}}{2} \right)^n$$
for all $n \in N$.

(c) Show that x_{3n} is even for all $n \in \mathbb{N}$.

(d) Conjecture and prove a theorem concerning the sum $x_1 + x_3 + x_5 + \cdots + x_{2n-3} + x_{2n-1}$.

(e) Conjecture and prove a theorem concerning the sum $x_2 + x_4 + x_6 + \cdots + x_{2n-2} + x_{2n}$.

(ii) Let $x_1 = 0$, $x_2 = 1$ and $x_n = \frac{1}{2}(x_{n-1} + x_{n-2})$. Show that

$$x_n = \frac{2^{n-1} + (-1)^n}{3 \times 2^{n-2}}$$

for all $n \in \mathbb{N}$.

(iii) Find the unique factorization of 12 870 and of 17 836.

Summary

▶ The initial step does not have to be $A(1)$. If $A(r)$ true and $A(k) \Longrightarrow A(k+1)$ for all $k \geq r$, then $A(n)$ is true for $n \geq r$.

▶ The inductive step does not have to be $A(k) \Longrightarrow A(k+1)$. We can have
  (i) $A(k-1)$ and $A(k)$ implies $A(k+1)$, or
  (ii) $A(j)$ true for all $j \leq k$ implies $A(k+1)$.

▶ We can combine changes to initial step and induction step. If $A(r)$ true and $A(j)$ true for all $r \leq j \leq k$ implies $A(k+1)$, then $A(n)$ is true for $n \geq r$.

CHAPTER 26

Techniques of proof V: Contrapositive method

If it isn't hurting, it isn't working.
Prime Minister John Major's comment after being told that his
government's policies were hurting the country

The contrapositive method is very confusing to grasp initially – however, one student told me that she liked using it as it made her feel that she had done something clever. I think this feeling comes from the fact that the method is very indirect.

We saw in Chapter 8 that the statement '$A \implies B$' is equivalent to 'not $B \implies$ not A'. The contrapositive method is the use of this equivalence. The indirectness and feeling of being clever is that in using it to prove '$A \implies B$' we start with not B (and proceed to show that not A is true).

Since this idea causes a lot of confusion for beginners we shall start with some revision of Chapter 8 and show, via a different proof to the one in that chapter, that a statement and its contrapositive are equivalent. We shall then see the contrapositive in action.

Revision of the contrapositive

Definition 26.1

*The **contrapositive** of the statement '$A \implies B$' is*

  '$\text{not}(B) \implies \text{not}(A)$'.

Examples 26.2

(i) 'I am Winston Churchill implies I am English' has contrapositive 'I am not English implies I am not Winston Churchill.'

(ii) 'If I am Jane, then I am a woman' has contrapositive 'If I am not a woman, then I am not Jane.'

(iii) 'If $x^2 - 9 = 0$, then $x = 2$' has contrapositive 'If $x \neq 2$, then $x^2 - 9 \neq 0$.'

The contrapositive should not be confused with the inverse statement, which is 'not $A \implies$ not B'. This is not equivalent to the original statement as the following example shows.

Example 26.3

The statement '$x = 2$ implies that x is even' is true. Its inverse '$x \neq 2$ implies x is odd' is not true. For instance, x can be 6.

A statement and its contrapositive are equivalent. For example, in (i) and (ii) you can see that both statement and contrapositive are true, and in (iii) both are false. Of course, showing that something holds for a few examples is not enough. We used truth tables in Chapter 8 to show they are equivalent in general.

However, you need to feel that this equivalence is really true. To get a better feel for it take the example 'I am Jane implies that I am a woman.'

Assume this is true and consider whether the contrapositive 'I am not a woman implies I am not Jane' is true or not.

Well, if I am not a woman, then there is absolutely no way that I am Jane since we know that Jane is a woman. Hence, the contrapositive 'If I am not a woman, then I am not Jane' is true.

More generally, suppose that '$A \Longrightarrow B$' is true. Then, the contrapositive is 'not $B \Longrightarrow$ not A'. To see it must also be true, then assume that not B is true. In this case, there is absolutely no way that A can be true, since if it was, then B would be true (because $A \Longrightarrow B$), so A must be false, i.e. not A is true.

Hence we have shown that

$$\text{`}A \implies B\text{'} \implies \text{`not } B \implies \text{not } A\text{'}. \qquad (*)$$

To show equivalence we need to show the converse is true. For this we can apply similar reasoning to the above but instead use a clever trick. Let C be the statement not A and D be the statement not B. Then if 'not $B \Longrightarrow$ not A' is true then '$D \Longrightarrow C$'. So from $(*)$ we know that 'not $C \Longrightarrow$ not D', i.e. not(not A)) $\Longrightarrow$ not(not(B)). This is just '$A \Longrightarrow B$'.

Let us summarize all this and write it out cleanly in a proof.

Theorem 26.4

The statement '$A \Longrightarrow B$' is equivalent to the statement 'not $B \Longrightarrow$ not A'.

Proof. As the theorem is a statement about equivalence it is an 'if and only if' type statement which we can break into an 'if' statement ($\Longleftarrow$) and an 'only if' statement ($\Longrightarrow$).

[$\Longrightarrow$] Suppose that $A \Longrightarrow B$ but assume (for a contradiction) that not $B \Longrightarrow$ not A is false. This assumption basically says that not B is true but not A is false. But if not A is false, then A is true, and as $A \Longrightarrow B$, we get that B is true, i.e. not B is false. This is a contradiction.

[$\Longleftarrow$] Suppose that not $B \Longrightarrow$ not A. By [$\Longrightarrow$] above this implies that not(not A) $\Longrightarrow$ not (not B). That is, $A \Longrightarrow B$. $\qquad \square$

Note that we removed any mention of C and D.

Another way of looking at $A \Longrightarrow B$ is to say that if A is true, then there is no way B is false. Thus if B is false then A can't be true. So we have not $B \Longrightarrow$ not A.

Using the contrapositive

In the proof of Theorem 18.1 we have already seen an example of a proof employing the contrapositive.

Example 26.5

We shall prove that

'If x^2 is even, then x is even'

by the contrapositive method. This is equivalent to proving

'If x is not even, then x^2 is not even.'

This in turn means that

'If x is odd, then x^2 is odd.'

This is easy to show: if x is odd, then $x = 2k + 1$ for some $k \in \mathbb{Z}$. We have

$$x^2 = (2k + 1)^2 = 4k^2 + 2k + 1 = 2(2k^2 + k) + 1,$$

which is odd.

We have shown directly that 'x odd implies that x^2 is odd.' Hence the contrapositive statement 'x^2 is even implies that x is even' is true.

Example 26.6

Suppose that A, B, C and D are sets such that $C \setminus D \subset A \cap B$ and that $x \in C$. Prove that if $x \notin A$, then $x \in D$.

To prove this we would probably first try to do it directly. Instead we will show the contrapositive: If $x \notin D$, then $x \in A$.

Therefore, let us suppose that $x \notin D$. Since $x \in C$ is assumed, then $x \in C \setminus D$. Because $C \setminus D \subset A \cap B$ this implies $x \in A \cap B$, i.e. $x \in A$.

In this last example the polished proof would begin with the last line beginning with 'Let us suppose that $x \notin D \ldots$' and would not make any mention that the contrapositive statement was being proved. It would be for the reader to realise that.

Don't confuse contradiction and contrapositive

Proof by contradiction for $A \implies B$ and proof by contrapositive do look similar and you can be forgiven for getting the two confused. Both prove statements of the form $A \implies B$ and both use negation in doing so. The difference is subtle and to complicate matters we used proof by contradiction to show that proof by contrapositive worked! So, what's the difference? Let's be clear about the methods:

Contradiction:	Assume A and not B are both true and show that a contradiction results.
Contrapositive:	Assume not B and show not A.

The contrapositive method has the advantage that you turn the problem into a direct problem, the goal is clear: you assume not B is true and you have to show that not A is true. In the contradiction method it may not be obvious what the contradiction is going to be; your goal is less clear.

Exercises

Exercises 26.7

(i) Suppose x and y are natural numbers. Show that xy odd implies that x and y are both odd.

(ii) Suppose that x and y are real numbers. Prove that if $x + y$ is irrational then x is irrational or y is irrational.

(iii) Show that if $x^2 - 3x + 2 < 0$ then $1 < x < 2$.

(iv) Prove that if x is irrational, then $\sqrt{x}$ is irrational.

(v) Consider the equation $ax^2 + bx + c = 0$ with $a, b, c \in \mathbb{Q}$. Prove that one solution is irrational if and only if the other is irrational.

(vi) Prove or give a counterexample to the statement: If x and y are irrational then x^y is irrational.

(vii) Consider the quote from Prime Minister John Major at the start of this chapter: If it isn't hurting, it isn't working. This was his answer to being told that his government's policies to get out of recession were hurting the country. What is wrong with his logic?

Summary

▶ The contrapositive statement of '$A \Longrightarrow B$' is 'not $B \Longrightarrow$ not A'.

▶ Don't confuse proof by contradiction and proof by contrapositive.

PART V

Mathematics that all good mathematicians need

CHAPTER 27

Divisors

You know my methods. Apply them.
Sherlock Holmes
in Arthur Conan Doyle, *The Hound of the Baskervilles*

The set of integer numbers is among the simplest objects in mathematics. These numbers have a whole theory devoted to them, called, not surprisingly, number theory. In this chapter we will consider a small part of this theory. The aim is to provide you with mathematics that you will often need but, more importantly, we will apply the earlier methods to see how they illuminate the material.

Divisibility

What basic properties do numbers have? What can I do with them? Well, I can add and subtract then, multiply and divide, take to the power just to name a few. The best set of numbers to begin playing with is the integers, denoted $\mathbb{Z}$. Adding any two integers produces an integer. Subtracting any integer from another produces an integer. Multiplying any two integers produces an integer. Dividing any integer by another produces an integer... sometimes. This means we have found an interesting property concerning division of integers. Being mathematicians we produce a definition to isolate this idea and start to investigate it.

Definition 27.1

*An integer a **divides** the integer b if there exists an integer k such that $b = ka$. In this case we say b is **divisible** by a and write $a|b$. We also say that a is a **divisor** of b. If a does not divide b, then we write $a \nmid b$.*

Remarks 27.2

(i) We have actually met this idea before. The concept of primes depends on divisibility by natural numbers and we used it in some induction examples in Chapter 24.
(ii) Note that, as usual, the definition uses the word 'if' but actually means 'if and only if'. So when $b = ka$ we say a divides b, and conversely when we say a divides b, then we know that there is a k such that $b = ka$.

Let's think like mathematicians as we analyze the definition: What examples are there? What non-examples are there?

Examples 27.3

(i)   We have 3|6 since $6 = 2 \times 3$, and 2|6 for the same reason.

(ii)   We also have $-3|6$ as $6 = (-2) \times (-3)$. Note that in the definition we are allowed integers; divisors do not have to be natural numbers.

(iii) Any even number is divisible by 2. (This is practically the definition of even!)

(iv)   146 552 442 374 356 965 490 is divisible by 10.

(v)   For any even integer n, the integer $3n$ is divisible by 6, because it is divisible by 2 and 3. I.e. n is even so $n = 2m$ for some m and so $3n = 3 \times 2m = 6m$. When we divide n by 6 we get an integer (i.e. m) so n is divisible by 6.

(vi)   5 does not divide 33.

Exercises 27.4

(i)   Create your own examples and counterexamples.

(ii)   How many different ways can you generalize the example about $6|3n$ for n even?

(iii) In that example, consider what happens when we change the hypothesis to n odd. What can we say?

Note that we have already generated a lot of questions. This is how real mathematicians work.

What are the trivial examples of this concept? It's about integers so the numbers most involved in 'trivial' situations are 0, 1 and -1. A little bit of thought shows that every integer divides 0 and that 1 and -1 divide everything. An extreme example is that a number divides itself. (You may view this as a trivial example if you wish!)

What happens when we add two numbers or, more generally, add multiples of two numbers which are divisible by a? The next theorem gives an answer.

Theorem 27.5

If $a|b$ and $a|c$, then $a|(mb + nc)$, for all integers m and n.

Now let us apply the ideas in Chapter 16, How to read a theorem. First, we should try a few examples to see that the statement holds and to get a feel of why it should be true. So, take $a = 3$, then we could have $b = 12$ and $c = 21$, since then the hypotheses are satisfied. Pick your own examples that satisfy the hypotheses! The theorem says we can pick *any* integers for m and n. I'll take $m = 4$ and $n = -6$. I deliberately took a negative number as it is tempting when dealing with integers to pick positive ones. I often find that picking negative ones provides more insight.

Thus, $mb + nc = (4 \times 12) + (-6 \times 21) = 48 - 126 = -78$. Is this divisible by a, i.e. 3? Well, $-78/3 = -26$. Thus $a|mn + bc$ as claimed by the theorem.

Of course, this only shows the theorem for a specific example – and one specific case does not prove the general! We still need to provide the general proof. Let's continue our explorations though.

What are the trivial situations for this theorem? We are allowed to take any integers m and n. Trivial examples of integers are 0, 1 and -1.

Let's try $m = 1$ and $n = 0$. From this we can see that if $a|b$ and $a|c$, then a divides $(1 \times b) + (0 \times c)$. But this latter number is b so we can deduce that a divides b. To recap we have shown

If $a|b$ and $a|c$, then $a|b$.

Hmm ... That's not a very good statement. It says 'If X and Y are true, then X is true.' Oh well, this happens in exploration.

Let's try another trivial example $m = n = 1$. This gives

If $a|b$ and $a|c$, then $a|(b + c)$.

Now this *is* good. It tells us something we could probably use later. That is, if a divides b and a divides c, then it divides their sum. Similarly, we take $m = 1, n = -1$ to get

If $a|b$ and $a|c$, then $a|(b - c)$.

The point here is that in investigating we have found something useful.

Exercise 27.6

Try applying other techniques from Chapter 16 to see what you can find.

We now prove the theorem. Initially, this is not presented in the way that a standard textbook presents the proof. I wish to show how a proof is created with mistakes and dead ends. Then I sum it all up in the way that a book would.

Let us apply some problem-solving techniques – ask the question 'What do we know?' and 'What does that mean/imply?'

What do we know? We know that $a|b$ and $a|c$. What does this mean? To know what it means we go back to the definition. (Recall from Chapter 15, How to read a definition, that it is vital that you know the definition precisely – vaguely is not enough.) Looking back to the definition (you did look if you needed to, didn't you?) we find that it means that there exist integers k_1 and k_2 such that $b = k_1a$ and $c = k_2a$.

What do I want to know? I want $a|(mb + nc)$ for all integers m and n. What does this mean? It means that there exists k_3 such that $mb + nc = k_3a$. My experience tells me that this k_3 should be dependent on k_1 and k_2, i.e. what I want depends on what I know. Let's do some calculating!

We know $b = k_1a$ and $c = k_2a$ and want $mb + nc = k_3a$. As usual we look at the complicated side of the equation and try to reduce:

$$mb + nc = m(k_1a) + n(k_2a), \text{ by straightforward substitution of assumptions,}$$
$$= (mk_1 + nk_2)a, \text{ by some simple algebra.}$$

This shows that $a|(mb + nc)$ for all m and n and the k_3 we seek is $k_3 = mk_1 + nk_2$.

We've done all the work; let's write this up properly. To indicate the polished versions of proofs – the type that would be submitted in an assignment – we will use a box.

Polished solution:

Proof. By assumption, there exist integers k_1 and k_2 such that $b = k_1 a$ and $c = k_2 a$. For any integers m and n we have

$$mb + nc = m(k_1 a) + n(k_2 a), \text{ by assumption,}$$
$$= (mk_1 + nk_2)a.$$

Thus, $mb + nc$ is divisible by a. □

This finishes the construction of the proof.

Remark 27.7

Lessons to be learned:

- The integer k_3 does not appear in the polished solution. It isn't necessary, all I needed was the realization that $(mk_1 + nk_2)a$ is divisible by a.
- Notice the questioning structure used in the search for a proof – What do I know? What does that mean? What do I want to know? What does that mean?

Now, let us look at a corollary of this result. You may have already discovered it when you did the analysis from Chapter 16.

Corollary 27.8

Let a and b be integers. If a divides b, then a divides b^2.

Proof. Take $m = b$ and $n = 0$ in Theorem 27.5. □

This is a perfect example where we can say 'What is the converse?' The converse is

'Let a and b be integers. If a divides b^2, then a divides b.'

It seems plausible, but it is false! For example, take $a = 4$ and $b = 6$. Then $b^2 = 6^2 = 36$ which is divisible by $a = 4$, but 4 does not divide 6.

We state two more simple facts.

Theorem 27.9

Let a, b and c be integers.

(i) *If $a|b$ and $b|c$, then $a|c$.*
(ii) *If $a|b$ and $b|a$, then $a = b$ or $a = -b$.*

We shall now explain in detail how to find the proof of each statement before writing the final polish version.

(i) How are we going to do this? The trick we have used – in fact, it is the definition – is to rewrite a number in terms of its factors. Let's try that again.

Well, $a|b$ and $b|c$ implies the existence of integers k_1 and k_2 such that $b = k_1 a$ and $c = k_2 b$. We can replace the b in the latter equation to get $c = k_2 k_1 a$. So, a divides c.

Polished solution:

Proof. By assumption, there exist integers k_1 and k_2 such that $b = k_1 a$ and $c = k_2 b$. Hence, $c = k_2 k_1 a$, and we deduce that a divides c. □

Remark 27.10

Lesson to be learned: Notice that we used the same trick as in the previous theorem – apply the definition. This again shows how important it is to learn definitions.

Consider now the proof of (ii). Look at the assumptions and their implications. That is, what do we know and what does it mean?

We know that $b = k_1 a$ and $a = k_2 b$. Thus $b = k_1 k_2 b$. This implies that $k_1 k_2 = 1$. We know that k_1 and k_2 are integers so they must be either 1 or -1. Thus $a = \pm b$ or $b = \pm a$. Wait, these are the same condition! So $a = \pm b$.

Exercise 27.11

Write a polished solution. Or give your own proof!

Exercises 27.12

(i)   Are the converses to the statements in the last theorem true? If so, give a proof, if not, then give a counterexample.

(ii) Prove that if $a|b$ and $c|d$, then $ac|bd$.

(iii) If $a|b$ and $c|b$, then does $ac|b$? If so, prove it. If not, give a counterexample.

Now suppose that the number we wish to divide is given by a polynomial.

Example 27.13

For n even, $n^2 + 2n + 8$ is divisible by 4.

To show this, we consider what it means for n to be even: $n = 2m$ for some integer m. Then, we look at the expression we wish to find is divisible by 4:

$$n^2 + 2n + 8 = (2m)^2 + 2(2m) + 8 = 4m^2 + 4m + 8 = 4(m^2 + m + 2).$$

Since $m^2 + m + 2$ is an integer we conclude that $n^2 + 2n + 8$ is divisible by 4.

Exercise 27.14

Show that $x^2 + 9x + 20$ is divisible by 2 for all $x \in \mathbb{Z}$.

There exist an infinite number of primes

The next theorem is another classic of mathematics. The first known proof was by Euclid.

Theorem 27.15

The number of prime numbers is infinite.

Proof. Assume the contrary statement: there is a finite number of primes, p_1 to p_n.

The number $p_1 p_2 \ldots p_n + 1$ is larger than all of p_1 to p_n so by assumption cannot be a prime. On the other hand, if we divide this number by p_j for any j, then we get remainder 1. Now suppose that this number is divisible by a non-prime less than it, say b. By the Fundamental Theorem of Arithmetic p_i divides b for some p_i and so p_i divides

$p_1 p_2 \dots p_n + 1$, which we have seen cannot be true. This means that $p_1 p_2 \dots p_n + 1$ is only divisible by itself and 1. That is, it is prime.

Hence, we have shown that if there is a finite number of primes, then $p_1 p_2 \dots p_n + 1$ is not a prime and is a prime. This is an obvious contradiction.   □

Exercise 27.16

Analyse the above proof. How many times did we use contradiction?

Greatest common divisor

Definition 27.17

*The **greatest common divisor** of two non-zero integers a and b, denoted $\gcd(a, b)$, is the largest positive integer that divides both numbers.*

*It is sometimes known as the **greatest common factor** (gcf) or **highest common factor** (hcf).*

Examples 27.18

(i)   The positive divisors of 20 are 1, 2, 4, 5, 10, and 20. The positive divisors of 12 are 1, 2, 3, 4, 6 and 12. Therefore the greatest common divisor is 4.

(ii) The greatest common divisor of 65 and 36 is 1.

Exercise 27.19

Create your own examples and non-examples. Include some with negative numbers, e.g. $\gcd(50, -35)$.

Let's look at some basic properties of the gcd.

Theorem 27.20

For any integers a and b we have

(i) $\gcd(a, b) = \gcd(b, a)$,

(ii) $\gcd(a, b) \geq 1$,

(iii) $\gcd(a, b) = \gcd(|a|, |b|)$,

(iv) $\gcd\left(\dfrac{a}{\gcd(a, b)}, \dfrac{b}{\gcd(a, b)}\right) = 1$,

(v) $\gcd(a, b) = \gcd(a + nb, b)$ *for all integers n.*

Proof. (i) This is trivial. We just use the definition.

(ii) The gcd is defined to be a positive integer so it must be greater than or equal to 1.

(iii) This is a simple exercise.

(iv) Let's begin by giving $\gcd(a, b)$ a name. Let $\gcd(a, b) = d$. So we have $a = rd$ and $b = sd$ for some integers r and s as the greatest common divisor is obviously a divisor. Hence,

$$\gcd\left(\frac{a}{\gcd(a, b)}, \frac{b}{\gcd(a, b)}\right) = \gcd\left(\frac{rd}{d}, \frac{sd}{d}\right) = \gcd(r, s).$$

Let's give this gcd a name too. Let $\gcd(r, s) = d'$. Then $r = pd'$ and $b = qd'$ for some p and q. So we know that $a = pd'd$ and $b = qd'd$. This though means that $d'd$ is a divisor of a and b. If d' is greater than 1, then this would contradict that d is the greatest common divisor of a and b as $d'd$ would be greater than d.

(v) This involves saying that the greatest common divisors of a, b and the sum $a + nb$ are connected, so in particular says something about their divisors. Hence I ask myself what theorems do I know that connect divisors and sums? By looking back I notice that Theorem 27.5 has assumptions and conclusions involving a divisor and sums so I will use this and see what happens.

We have $\gcd(a, b)|a$ and $\gcd(a, b)|b$ by definition of gcd. So $\gcd(a, b)$ divides $a + nb$ by Theorem 27.5. Hence $\gcd(a, b)$ is a common divisor for b and $a + nb$. It may not be the greatest though, therefore we have $\gcd(a, b) \leq \gcd(a + nb, b)$.

This is not what we wanted to prove! We wanted an equality. It's not a problem though since $x = y$ is equivalent to $x \leq y$ and $x \geq y$. Thus we need to prove $\gcd(a, b) \leq \gcd(a + nb, b)$. It seems reasonable that we can use an argument similar to the above. The proof is almost identical:

We have $\gcd(a + nb, b)|a + nb$ and $\gcd(a + nb, b)|b$ by definition of gcd. So $\gcd(a + nb, b)$ divides $(a + nb) + (-n)b$ by Theorem 27.5. Hence $\gcd(a + nb, b)$ is a common divisor for b and a. It may not be the greatest, therefore $\gcd(a + nb, b) \leq \gcd(a, b)$.

Hence, as $\gcd(a, b) \leq \gcd(a + nb, b)$ and $\gcd(a + nb, b) \leq \gcd(a, b)$, we have $\gcd(a + nb, b) = \gcd(a, b)$. $\qquad\square$

Exercises 27.21

(i)   Write a polished proof for (iv) as a proof by contradiction.

(ii)   Write a polished version of the proof of (v). Can you combine the two (almost) identical arguments?

(iii) What lessons have you learned form this theorem and proof?

A common mistake

A common error made by beginners is that if $n|ab$, then n must divide one of a or b. This is not the case.

Example 27.22

Let $n = 6$, $a = 4$ and $b = 9$. Then $n|ab$ is $6|36$ which is true. But $6 \nmid 4$ and $6 \nmid 9$.

This example is a counterexample to the statement and if we look further we can see what happens. We have $a = 4 = 2 \times 2$ and $b = 9 = 3 \times 3$ so $ab = (2 \times 2 \times 3 \times 3) = 2 \times (2 \times 3) \times 3$. Thus the divisor, 6, arises partly from a and partly from b.

We shall see later (in Euclid's Lemma, page 203) that we can add an extra condition to the above erroneous statement to produce a theorem. For the moment it is best to be aware of this misunderstanding.

Exercises

Exercises 27.23

(i)   Consider the following statements
    (a) n is divisible by 3,
    (b) n is divisible by 9,
    (c) n is divisible by 12,
    (d) $n = 24$,
    (e) n^2 is divisible by 3,
    (f) n is even and divisible by 3.
    Which conditions are necessary for the natural number n to be divisible by 6? Which are sufficient? Which are necessary and sufficient?

(ii)   For each positive integer, show that $x^3 - x$ is divisible by 3 and $x^5 - x$ is divisible by 5. Can you generalize this? Is $x^n - x$ divisible by n?

(iii)   Show that $x^3 - 6x^2 + 11x - 6$ is divisible by 3 for all $x \in \mathbb{Z}$.

(iv)   Show that for $n \in \mathbb{N}$ if $n \geq 4$ is not prime, then $n|(n-1)!$.

(v)   Prove that every common divisor of a and b is a divisor of $\gcd(a, b)$ for all $a, b \in \mathbb{Z}$.

(vi)   Prove $\gcd(m+1, n+1)|mn - 1$ for all $m, n \in \mathbb{Z}$.

(vii)   A common error (given above) is that 'if $n|ab$, then n must divide one of a or b'. This is not true. If we add the assumption that $a = b$ what statement do we get and is it true? What happens if, furthermore, we assume n is prime?

(viii)   Prove that $m^2|n^2 \implies m|n$ for all $m, n \in \mathbb{Z}$.

(ix)   Prove the following.
    (a) A natural number is divisible by 2 if and only its last digit is divisible by 2.
    (b) A natural number is divisible by 4 if and only the number formed by its last two digits is divisible by 4.
    (c) A natural number is divisible by 8 if and only the number formed by its last three digits is divisible by 8.
    Can you generalize this and prove your generalization?

(x)   Prove that $x - y$ divides $x^n - y^n$ for all $n \in \mathbb{N}$ and all $x, y \in \mathbb{Z}$. Hence show that
    (a) $4 \times 11^n + 2 \times 5^n$ is divisible by 6 for all n,
    (b) $6 \times 9^n - 4^n$ is divisible by 5 for all n.

(xi)   Define the nth **Fermat number** to be $F_n = 2^{2^n} + 1$ where $n \in \mathbb{N}$.
    (a) Prove that

$$\prod_{k=0}^{n} F_k = F_{n+1} - 2.$$

    (b) Hence, show that for every pair of distinct Fermat numbers F_j and F_l we have $\gcd(F_j, F_l) = 1$.

Summary

▶ An integer a divides the integer b if there exists an integer k such that $b = ka$.
▶ We say b is divisible by a and write $a|b$.

▶ We say that a is a divisor of b.

▶ If $a|b$ and $a|c$, then $a|(mb + nc)$, for all integers m and n.

▶ The greatest common divisor of two non-zero integers a and b, denoted $\gcd(a, b)$, is the largest positive integer that divides a and b.

▶ $n|ab$ does *not* imply that n divides one of a and b.

CHAPTER 28

The Euclidean Algorithm

An algorithm must be seen to be believed.
Donald Knuth, *The Art of Computer Programming*, Vol. 1, 1999

The Euclidean Algorithm[1] is a very powerful device. We will apply it to three problems:

- Finding the gcd of two numbers.
- Finding *integer* solutions to equations like $32x + 17y = 45$, i.e. those of the form $ax + by = c$.
- Finding additional hypotheses so that when $n|ab$ we can say something about whether or not n divides a or b.

We will show how the proofs are created and how they are polished for the final version. Hopefully, it will show that there is a huge difference between the way a proof is created and how it is presented.

The Division Lemma

The obvious way to find the gcd of two integers is to factorize them – we know we can do this by the Fundamental Theorem of Arithmetic (Theorem 25.5) – and find what common prime factors they have. Multiplying these together gives the gcd. For example, $440 = 2^3 \times 5 \times 11$ and $1300 = 2^2 \times 5^2 \times 13$. The common factors are 2^2 and 5 so the gcd is $2^2 \times 5 = 20$.

The problem is that factorization is hard for large numbers.[2] Instead of this brute force method we shall give a more useful method. We start with a lemma.

Lemma 28.1 (Division Lemma)

Let x and y be non-zero integers with $y > 0$. Then, there exist unique integers q and r such that $x = qy + r$ where $0 \le r < y$.

You should of course convince yourself that this statement is reasonable. Try a few examples. It is easy to see what q and r could be. Divide x by y and you will get a number

---

[1] An algorithm is just a list of instructions for completing a specific task.
[2] And I mean hard! Security on the internet depends on this fact.

which probably isn't an integer. Take the integer part, call it q. Then set $r = x - qy$. This gives the right q and r.

You should be asking Why do we need non-zero integers? Why do we need $y > 0$? (These will be answered later.)

Proof (of the Division Lemma). Let's get into mathematical thinking mode. The conclusion of the statement is that there exist certain unique numbers. The standard way to approach such a problem is to prove the existence and later prove uniqueness. (Put this idea into your toolbox of ideas for tackling problems.) We also have to prove a property ($0 \leq r < y$) holds – maybe this will follow from existence.

It should be obvious that we are trying to divide x by y and get the remainder r. The easiest way to do division and to avoid all the horrible mathematics involved in dealing with real numbers rather than integers is to do subtraction. That is, we look at $x - y$, $x - 2y$, $x - 3y$ and so on until we get a number that y cannot be subtracted from without being negative; that number is then r. Hence I am looking for the smallest integer of the form $x - sy \geq 0$ where s is an integer. Call the least element r and let q be the integer such that $r = x - qy$. Then q and r are the integers we seek as $r = x - qy$ can be rearranged to $x = qy + r$.

Consider the set $S = \{x - sy \mid s \in \mathbb{Z} \text{ and } x - sy \geq 0\}$. Its smallest element is the r we want. Ok, we need to show that it has a smallest element. It is obvious that a non-empty set of positive integers must contain a smallest element.[3] Therefore all I have to do is check that the set is non-empty – I don't have to *find* the smallest element; I just need it to exist!

If $x \geq 0$, then $x \in S$ as $x - (0 \times y) = x \in S$. What if $x < 0$? In this case we need s negative so that $x - sy \geq 0$. Well, if $x < 0$, then $x - xy = x(1 - y) \geq 0$. Therefore $x(1 - y) \in S$.

Therefore, S is a non-empty set of positive integers and so it must have a smallest element. And, as we have seen, this gives us our r and q.

Now we'll rewrite this whole argument in a mathematical way. Note that our working out above is totally different to the polished answer – for example we start the polished version by defining the set S without explaining why we need it or where it comes from. That's ok as we're trying to convince someone that q and r exist – we are not trying to show them how we *found* the proof.

Polished solution:
Consider the set $S := \{x - sy \mid s \in \mathbb{Z} \text{ and } x - sy \geq 0\}$. If $x < 0$, then $x - xy = x(1 - y) \geq 0$. Therefore, $x(1 - y) \in S$. If $x \geq 0$, then $x - (0 \times y) = x \in S$. Hence, S is a non-empty set of positive integers and so has a smallest element. Call this smallest element r and let q be the integer such that $r = x - qy$. Then q and r are the integers we seek as $r = x - qy$ can be rearranged to $x = qy + r$.

Lesson learned: Notice that the polished proof starts somewhere in the middle of our original working. We begin by defining the set S. I do not believe this would be anyone's

---

[3] I say obvious but really I am hiding something – I'll be in so much trouble from my colleagues for stating it was obvious. I'll get letters, emails, death threats, etc. if I don't mention that it follows from the **Well-ordering Principle**. See the exercises at the end of the chapter; it's not important at the moment.

first step when trying to come up with a proof. Hence, the lesson is that the proof and the working are extremely different.

That's the existence problem sorted. However, we were required to show that $0 \leq r < y$. The first inequality is obvious as r is from S, a set of non-negative integers. How do we show $r < y$? We do what a mathematician would do. To compare two numbers we subtract one from the other and see what happens:

$$r - y = (x - qy) - y = x - (q + 1)y.$$

Now, we know that $r - y$ is in the right '$x - sy$' form to be in S, but it's less than r (can you see why?) contradicting the fact that r is the smallest element in S. Hence, $r - y$ must fail the other condition to be in S (that $x - sy \geq 0$), and so we must have $r - y = x - (q + 1)y < 0$. Therefore, $r < y$.

Lesson learned: To show $r < y$ we look at $r - y$ and show that this is less than 0.

Let's make a polished version of this. This follows on directly from the last polished answer.

> **Polished solution:**
> As $r \in S$ and S is a set of non-negative integers we have $0 \leq r$. We also have
>
> $$r - y = (x - qy) - y = x - (q + 1)y,$$
>
> so $r - y$ will be in S if $x - (q + 1)y \geq 0$. But r is the smallest member of S and $r - y$ must be smaller. Hence, $r - y = x - (q + 1)y < 0$. That is $r < y$.

To show uniqueness we do what mathematicians do and use contradiction. We assume that the numbers are *not* unique and show that this leads to a contradiction. That is, we assume they are different and then show that their difference is zero, in other words, they are not different. They cannot be different and not different at the same time!

Hence, suppose that there exist q' and r' which satisfy $x = q'y + r'$ with $0 \leq r' < y$, and $q \neq q'$ or $r \neq r$. (This latter means that the pairs (q, r) and (q', r') are different.) Then, we do the usual process of subtraction to get

$$0 = (q - q')y + (r - r').$$

This implies $(q - q')y = r' - r$. It is trivial that if $q = q'$, then $r' = r$. This would contradict that the pairs (q, r) and (q', r') are different.

Thus we must have $q \neq q'$. Let us consider the case $q - q' > 0$ as this gives us $(q - q')y$ positive in $(q - q')y = r' - r$. I'm not sure that this is the right way to go; it just makes it easier. If $q - q' > 0$, then $q - q'$ is at least 1, so $y \leq r' - r$. But this is a contradiction as r and r' are non-negative and less than y.

This deals with the case $q > q'$. Now, we can probably carry out a similar argument for $q < q'$, i.e. $q - q' < 0$, but notice that there was nothing special about the q and q'. Without loss of generality I can assume q is bigger than q' all the time. Because if it wasn't, then I could rename q' as q, and q as q'.

Now for the polished version.

Polished solution:

Suppose that q and r are not unique, so there exist q' and r' which satisfy $x = q'y + r'$ with $0 \leq r' < y$ and $q \neq q'$ or $r \neq r'$. If $q = q'$, then $r = x - qy = x - q'y = r'$. Hence, we can assume that q and q' are different. Without loss of generality, we can assume that $q > q'$. Then, $x = qy + r$ and $x = q'y + r'$ imply that $(q - q')y = r' - r$. We have $q - q' > 0$ and in particular $q - q'$ must be at least 1 so $y \leq r' - r$. But this is a contradiction as r and r' are non-negative and less than y.

As you can see we have done all the parts: existence (including the condition to be satisfied) and uniqueness.   □

More general version of the Division Lemma

Let us look at the assumptions of the Division Lemma again. Was the $y > 0$ necessary? After all it seems reasonable that if $y = -5$ and $x = 21$, then we should be able to find integers q and r: $21 = (-4) \times (-5) + 1$. Did we use actually use the assumption that $y > 0$? In fact, we did. If y were negative, then for positive s we would be adding in the expression $x - sy$, not subtracting as required.

So can we drop the $y > 0$ assumption? It was necessary in the proof, but maybe there is an alternative proof that allows us to drop it. The answer is yes, provided we change the conclusion slightly.

The $y > 0$ assumption is needed in the $0 \leq r < y$ condition. If $y = -5$ say, then we require $0 \leq r < -5$, a condition that can never be fulfilled.

What we do instead is take the modulus of y. Then, the condition for $y = -5$ is $0 \leq r < |-5| = 5$. In fact, there is a theorem:

Lemma 28.2 (a more general Division Lemma)

Let x and y be non-zero integers. Then, there exist unique integers q and r such that $x = qy + r$ where $0 \leq r < |y|$.

Proof. We have already proved the theorem in the case that $y > 0$ so we need to do the $y < 0$. Often when dealing with proofs involving cases we can repeat the proof with some minor modifications. In this case we probably can but would need to look at the largest element in S etc. However, in this case there is a clever trick for generalizing from the naturals to the integers. Furthermore, it can be used in other situations. Let's see this trick in action.

We know that the theorem holds for $y > 0$, so if we take $y' = -y$, then since y' is the negative of a negative, it must be positive. Thus, we can apply the Division Lemma to x and y':

There exist q' and r' such that $x = q'y' + r'$ and $0 \leq r < y'$. What does this tell us about what happens for y? Replacing y' by $-y$ we find that

$$x = q'(-y) + r' \quad \text{where } 0 \leq r' < -y.$$

This means that $x = (-q')y + r \quad \text{where } 0 \leq r' < |y|.$

Let $q = -q'$ and $r = r'$. We have found q and r such that $x = qy + r$ and have $0 \le r < |y|$.

Does this mean we have proved the new version? No, since the statement asks for *unique* q and r. When we proved uniqueness in the $y > 0$ case did we use $y > 0$? If you go back and check, then you can see that we didn't – it was used in the existence part. Hence, we can do one of two things. We can rewrite that part of the proof or we can say 'The proof of uniqueness is the same as in the proof of Theorem 28.1.'

Let's write out the polished proof of the general version of the lemma.

> Polished solution:
> **Proof (of a more general Division Lemma).** Let $y' = -y$. Then by Theorem 28.1 there exist integers q and r such that $x = qy + r$ with $0 \le r < y'$. Thus, $x = (-q)y + r$ with $0 \le r < |y|$. The proof of uniqueness is the same as in the proof of Theorem 28.1. □

Note that in the proof we did not use the notation q' and r', we just used q and r. The point is that the reader realizes that, when $x = (-q)y + r$, the $-q$ is the q we require! Notation can be confusing sometimes. □

Lessons learned:

- We stated two different versions of the Division Lemma, one more general than the other. If you check different books, you will see that authors usually just state one of them. It is quite common to see this in mathematics. Some authors don't need a result in its full generality and so make a choice and opt for a simpler statement.
- For problems involving integers try treating positive integers first. We may then get something that can be tackled – for example, we may be able to use induction. Then we apply the trick of using the positive case to prove the negative case.

The Euclidean Algorithm

The Euclidean Algorithm is just repeated use of the Division Lemma. In the lemma we put in x and y to produce q and r with $0 \le r < y$. We then put y and r into the lemma. We keep doing this until the remainder produced is zero.

Theorem 28.3 (Euclidean Algorithm)

Let x and y be integers. Then there exist integers $q_1, \ldots, q_k$ and a descending sequence of positive integers, $r_1, \ldots, r_k, r_{k+1} = 0$, such that:

$$x = q_1 y + r_1$$
$$y = q_2 r_1 + r_2$$

$$r_1 = q_3 r_2 + r_3$$
$$r_2 = q_4 r_3 + r_4$$
$$\vdots$$
$$r_{k-2} = q_{k-1} r_{k-1} + r_k$$
$$r_{k-1} = q_k r_k + 0.$$

Furthermore, $\gcd(x, y) = r_k$.

Here a descending sequence means that $r_1 > r_2 > r_3 > \cdots > r_k > 0$.

Remark 28.4

It is actually the last part of the statement that is the important part because it tells us how to calculate some number of interest, i.e. the gcd of two integers.

Proof. Evidently we apply the Division Lemma repeatedly. At each stage we have $0 \le r_i < r_{i-1}$. (Do you see why?) As we have a descending sequence of non-negative integers the process will produce a remainder of 0 at some point.

That $\gcd(x, y) = r_k$ follows from $\gcd(x, y) = \gcd(r, y) = \gcd(y, r)$ – which is Theorem 27.20 (i) and (v) – and from $\gcd(a, b) = a$ if $a|b$ – which is easy to see.   □

We shall now give some applications of the algorithm. The first application will give us a chance to see the algorithm in action.

Calculating gcd

Example 28.5

Find $\gcd(14\,441, 3563)$.

Repeatedly applying the Division Lemma produces:

$$14\,441 = 4 \times 3563 + 189$$
$$3563 = 18 \times 189 + 161$$
$$189 = 1 \times 161 + 28$$
$$161 = 5 \times 28 + 21$$
$$28 = 1 \times 21 + 7$$
$$21 = 3 \times 7 + 0.$$

Thus, $\gcd(14\,441, 3563) = 7$.

Notice that the method is just heads-down-no-nonsense mindless calculation. If we wanted to find the gcd by brute force we could find all the factors of $14\,441$ and 3563 – not an easy task – and multiply all the common factors to find the greatest common divisor. Try it and see that the above method is easier.

Exercises 28.6

(i)   Find $\gcd(315, 462)$.

(ii)   Find gcd(546, 9100).

(iii)   Without using Theorem 27.20(v) prove the following. Suppose that $x = qy + r$ for $x, y, q, r \in \mathbb{Z}$ where y does not divide x. Prove that $\gcd(x, y) = \gcd(y, r)$. (This is a perfect example of how to solve a problem by breaking it into pieces. First show that if $d = \gcd(x, y)$, then d is a divisor of y and r. Second, show that d is the greatest such integer. This last part is itself a perfect example of a proof by contradiction: assume that there is a greater one and proceed.)

By reversing the algorithm we can produce an interesting theorem.

Theorem 28.7

Suppose x and y are integers. Then, there exist integers k and l such that

$$kx + ly = \gcd(x, y).$$

Proof. Exercise (but look at the following example before attempting).   □

Example 28.8

Express $\gcd(14\,441, 3563)$ in the form $14\,441k + 3563l$ where $k, l \in \mathbb{Z}$.

To find this required expression we take the output of the Euclidean Algorithm in Example 28.5 and work backwards. In that example we found that $\gcd(14\,441, 3563) = 7$. We use the bottom line of our calculations first:

$$
\begin{aligned}
7 &= 28 - 1 \times 21 \\
&= 28 - 1 \times (161 - 5 \times 28), \text{ replaced 21 using 4th line of algorithm output,} \\
&= 6 \times 28 - 1 \times 161 \\
&= 6 \times (189 - 1 \times 161) - 1 \times 161, \text{ replaced 28 using 3rd line,} \\
&= 6 \times 189 - 7 \times 161 \\
&= 6 \times 189 - 7 \times (3563 - 18 \times 189), \text{ replaced 161 using 2nd line,} \\
&= 132 \times 189 - 7 \times 3563 \\
&= 132 \times (14\,441 - 4 \times 3563) - 7 \times 3563, \text{ replaced 189 using 1st line,} \\
&= 132 \times 14\,441 - 535 \times 3563.
\end{aligned}
$$

Thus, $k = 132$ and $l = -535$.

Bonus points to you if you asked 'Are the numbers k and l unique?' Double bonus points if you worked out that they weren't. It is not difficult to show that:

$$7 = (132 + 3563) \times 14\,441 - (535 + 14\,441) \times 3563.$$

Notice that we just used the trick that $ab - ba = 0$.

In general this lack of uniqueness is not a problem and will be used in the section on Diophantine equations.

Euclid's Lemma

Consider Example 27.22 where we saw that if $n|ab$, then we do not necessarily have that n divides one of a or b. By adding an extra hypothesis we can get such a statement. The resulting statement is called Euclid's Lemma. Note that the *lemma* is a *corollary* for us; it is a corollary of the Euclidean Algorithm!

Corollary 28.9 (Euclid's Lemma)

Suppose that n, a and b are natural numbers. If $n|ab$ and $\gcd(n, a) = 1$, then $n|b$.

Proof. Since $\gcd(n, a) = 1$ there exist integers k and l such that $kn + la = 1$. Thus $knb + lab = b$. We obviously have $n|knb$. We also have $n|ab$ so $n|lab$. Thus $n|knb + lab$, i.e. $n|b$. □

The lemma says that if n divides ab but doesn't divide a, then it must divide b.

The condition that the greatest common divisor of two numbers is 1 used in the lemma is so important that we give it a name.

Definition 28.10

*Two integers are called **coprime** or **relatively prime** if their greatest common divisor is 1.*

Exercise 28.11

Look for your own examples and non-examples of coprime integers.

From Euclid's Lemma we can deduce the following.

Theorem 28.12

If p is prime and p divides ab, then p divides a or p divides b.

Exercise 28.13

Prove the above theorem.

From this theorem we can prove a result on irrationality.

Corollary 28.14

If n is not a square number, then $\sqrt{n}$ is irrational.

Proof. Suppose to the contrary that $\sqrt{n} = r/s$, where r and s are integers. We can assume that r/s is in its lowest terms, i.e. after cancellation of any common factors, we may assume that $\gcd(r, s) = 1$. Since n is not a square, $s > 1$. We have $r^2 = ns^2$. Let p be any divisor of s that is prime. Then $p|r^2$, so $p|r$ by the previous theorem. Therefore, p divides both r and s, contradicting $\gcd(r, s) = 1$. □

Note that using divisibility results we have just proved something that on the face of it does not seem to be about divisibility.

Diophantine equations

Our next application concerns finding integer solutions of linear equations in two variables. We seek integer solutions to equations of the form $mx + ny = c$ where $m, n, c \in \mathbb{N}$ are given. We shall also get solutions to $mx - ny = c$ as a bonus. (Can you see why?) Such equations are called Diophantine equations (named after Diophantus of Alexandria).

Theorem 28.15

(i) *For all m and n in $\mathbb{N}$ there are integer solutions x and y to the equation $mx + ny = c$ if and only if $\gcd(m, n) | c$.*

(ii) *Suppose $x = X$ and $y = Y$ is a solution to $mx + ny = c$. Then, for all $t \in \mathbb{Z}$*

$$x = X + \frac{n}{\gcd(m, n)} t, \qquad y = Y - \frac{m}{\gcd(m, n)} t$$

is also a solution. Furthermore, all solutions are of this form.

Proof. I'll write out the ideas, you can write the polished solution as an exercise.

(i) This is an 'if and only if' statement so we break it down into an 'if' and an 'only if'. Obviously we are going to have to write out $\gcd(m, n)$ excessively during this question so let us declare that $d = \gcd(m, n)$ to reduce the writing.

Suppose there is a solution to $mx + ny = c$, say $mX + nY = c$. What do we want to show? We want to show that d divides c. What do we know? Well, we know c in terms of m and n (and X and Y). We know by definition that d is a divisor of m and n. By Theorem 27.5 a linear combination of m and n is also divisible by d. Thus, as $c = mX + nY$ (a linear combination) we know that d divides c.

Now let's go the other way in our equivalence. Suppose that d divides c. What do we know? In particular, what do we know about equations and greatest common divisors? We know that we can solve $mx + ny = d$ by Theorem 28.7. That is, we can find k and l such that $mk + nl = d$. (Note that the notation in that theorem clashes with the notation in this theorem – the ks and ls are not the same. No matter! We are mathematicians and we can cope with it.) But $mx + ny = d$ is not the equation we want to solve; we want c not d on the right-hand side. We can get that if we multiply d by c/d. What we do to one side we must do to the other. So, if we divide both sides by d and multiply by c, then we get

$$\frac{c(mk + nl)}{d} = c.$$

Thus, if d divides ck and cl, then we get integer solutions to $mx + ny = c$. That is, we want $ck/d \in \mathbb{Z}$ and $cl/d \in \mathbb{Z}$ in

$$m\frac{ck}{d} + n\frac{cl}{d} = c.$$

But this is ok as we know that d divides c by assumption.

(ii) Showing that the proposed solutions (given the solution with $x = X$ and $y = Y$) are actually solutions is merely a matter of substitution (put in $X + nt/\gcd(m, n)$ and $Y - mt/\gcd(m, n)$) as you can see for yourself. The main part of the statement is the 'furthermore' part, i.e. all solutions have that form.

Let us explore further. Suppose that all solutions for x have the form $X + \dfrac{n}{d}t$ for some $t \in \mathbb{Z}$. Does that give us the right form for y? It does; try it! That will be useful later and during the writing-up process. It should not be written at this stage. It should come at the end!

What do we know? We know that $mx + ny = c$ and $mX + nY = c$. What do we want? We want $x = X + \dfrac{n}{d}t$, etc. This doesn't have c in it and obviously we can eliminate c from the two equations we know. Thus we get the equation $m(x - X) + n(y - Y) = 0$, which is good because we want $x - X = \dfrac{n}{d}$. This last fact tells us that we need to involve d and so we divide the equation by d:

$$\frac{m(x - X)}{d} + \frac{n(y - Y)}{d} = 0.$$

Or in other words:

$$x - X = \frac{n}{d}\left(-\frac{d(y - Y)}{m}\right).$$

Hence, if we can prove that m divides $d(y - Y)$, then our integer t should be $-\dfrac{d(y - Y)}{m}$. That isn't too hard so I'll leave that to you to work out. If you get stuck, try the case that $\gcd(m, n) = 1$. $\qquad \square$

Exercise 28.16

Write a polished version of this proof.

Notice that part (ii) of the theorem says that if any solutions exist, then there are infinitely many of them.

Example 28.17

Find all the integer solutions to the Diophantine equation $51x + 21y = 18$.

First we find the greatest common divisor of 51 and 21:

$$51 = 2 \times 21 + 9$$
$$21 = 2 \times 9 + 3$$
$$9 = 3 \times 3 + 0.$$

Thus, the greatest common divisor is 3. Since $3 | 18$ we know that solutions exist. We can reverse the algorithm to find these:

$$\begin{aligned}
3 &= 21 - 2 \times 9 \\
&= 21 - 2(51 - 2 \times 21) \\
&= 5 \times 21 - 2 \times 51 \\
18 &= 30 \times 21 - 12 \times 51 \text{ after multiplying by } 18/3 = 6.
\end{aligned}$$

So,

$$x = -12 \text{ and } y = 30$$

is a solution to $51x + 21y = 18$. Hence (taking $X = -12$ and $Y = 30$) the solutions are of the form

$$X + \frac{21}{3}t = -12 + 7t, \quad Y - \frac{51}{3}t = 30 - 17t,$$

where t is any integer. That is,

$$x = -12 + 7t, \quad y = 30 - 17t$$

are the integer solutions to the equation $51x + 21y = 18$.

Example 28.18

We can also use these solutions to find solutions of $51x - 21y = 18$. (This is just the previous equation with the $+$ sign changed to a $-$ sign.) Since we can rewrite the equation as $51x + 21(-y) = 18$ we can see from the previous example that the solutions should be

$$x = -12 + 7t, \quad -y = 30 - 17t.$$

That is,

$$x = -12 + 7t, \quad y = -30 + 17t.$$

Exercises

Exercises 28.19

(i) Find (a) $\gcd(14\,592, 6468)$ (b) $\gcd(-12\,870, 4914)$.

(ii) Find integers m and n such that $14\,592m + 6468n = 12$.

(iii) Find integers m and n such that $4914m + 12\,870n = 234$.

(iv) Find all integer solutions to the Diophantine equations
   (a) $315x + 264y = 18$,
   (b) $315x + 264y = 24$,
   (c) $7644x + 1386y = 84$,
   (d) $1386x + 7644y = 126$.

(v) The **Well-ordering Principle** was used in the Division Lemma. It is an assumption we make about the natural numbers: every non-empty set of natural numbers has a smallest element.
   (a) Write the principle in the form 'If ..., then ...'
   (b) Can the hypothesis that the set is the natural numbers be generalized to the set of integers? If so, give proof; if not, give a counterexample. (By the way, it's not true for real numbers: for example, let $S = \{x \in \mathbb{R} \mid x > 0\}$.)
   (c) Where did we use this principle in the proof of the Principle of Mathematical Induction?

(vi) Prove the Division Lemma using strong induction on the integer x. (Hint: There are two cases to consider: $x < y$ and $x \geq y$.)

(vii) Consider Euclid's Lemma and its proof.
   (a) Identify in the proof where the hypotheses were used.
   (b) Which theorems were used in the proof?

(c) Analyse what happens when we drop the hypothesis that $\gcd(x, y) = 1$ in Euclid's Lemma. That is, drop $x, y \in \mathbb{N}$ to $x, y \in \mathbb{Z}$.

(viii) Assume that a and b are coprime. Prove that $\gcd(a + b, a - b) \leq 2$. (Hint: Let $g = \gcd(a + b, a - b)$, show that $g | 2a$ and $g | 2b$ and use cases on $\gcd(g, 2)$.)

(ix) In the more general version of the Division Lemma why doesn't uniqueness of q and r follow automatically from q' and r'?

(x) Let p_1 and p_2 be two numbers. Define q_i for $i \geq 1$ and p_i for $i \geq 3$ using the Division Lemma. That is, $p_{i-2} = q_{i-2}p_{i-1} + p_i$. Prove that $p_1 p_2 = \sum_{i=1} q_i p_{i+1}^2$. Draw a picture to explain this result.[4]

(xi) The **least common multiple** of two integers a and b, denoted $\operatorname{lcm}(a, b)$, is the smallest non-negative integer l such that there exist integers m and n such that $na = mb = l$. Prove the following.

(a) Any common multiple of a and b is a multiple of l.

(b) $\operatorname{lcm}(ra, rb) = r\operatorname{lcm}(a, b)$ for any positive integer r.

(c) $\operatorname{lcm}(a, b) \gcd(a, b) = |ab|$.

Summary

▶ The general division lemma: Let x and y be non-zero integers. Then there exist unique integers q and r such that $x = qy + r$ where $0 \leq r < |y|$.

▶ The Euclidean Algorithm can be used to calculate $\gcd(x, y)$.

▶ Euclid's Lemma: Suppose that n, a and b are natural numbers. If $n | ab$ and $\gcd(n, a) = 1$, then $n | b$.

▶ Two numbers are called coprime or relatively prime if their greatest common divisor is 1.

▶ For all m and n in $\mathbb{N}$ there are integer solutions x and y to the equation $mx + ny = c$ if and only if $\gcd(m, n) | c$.

▶ Furthermore, given one solution we can find all solutions.

▶ The least common multiple of two integers a and b, denoted $\operatorname{lcm}(a, b)$, is the smallest non-negative integer l such that there exist integers m and n such that $na = mb = l$.

---

[4] Lecturers: If you have never drawn a picture of this before, then do so. It is very revealing.

CHAPTER 29

Modular arithmetic

This is the kind of arithmetic I like.
Chico in *The Magnificent Seven*

When I was a student I was not particularly impressed with modular arithmetic. Initially, to me, it seemed like an abstract toy to play with. It didn't look as though it was useful or deep. Now I fully appreciate it as I have seen it used repeatedly, and widely, in real-life applications such as cryptography (which includes internet security), data storage, bar-codes, ISBN details (on the backs of books) and even magic tricks.

The idea of modular arithmetic is very simple. Take a natural number, say n. Then, we consider two integers to be equivalent if their remainders on division by n are equal. For example, take $n = 5$. Then 22 and 57 are the same because they both have the same remainder, in this case 2, when we divide by 5.

In this chapter we will study some elementary properties of modular arithmetic before moving on to some applications in number theory.

Modular arithmetic

We begin with the definition.

Definition 29.1

*If two integers x and y have the same remainder when we divide by a natural number n, then we say that x and y are **equivalent modulo** n (or x equals y mod n or x equals y modulo n) and write x mod $n = y$ mod n or more briefly $x = y$ mod n.*

Examples 29.2

(i) $15 = 0 \bmod 5$ because $15 = 3 \times 5 + 0$.
(ii) $16 = 4 \bmod 12$ because $16 = 1 \times 12 + 4$.
(iii) $-73 = 27 \bmod 10$ because $-73 = (-8) \times 10 + 7$ and $27 = 2 \times 10 + 7$.
(iv) $20 = 53 \bmod 11$ because both have remainder 9 after division by 11.

Obviously, if x has remainder r when we divide by n, then $x = r \bmod n$. In fact when asked for $x \bmod n$ we usually just give the remainder r. This is easy to find using a calculator.

Example 29.3

What is $261 \bmod 8$? We calculate that $261/8 = 32.625$. Thus, the remainder is $0.625 \times 8 = 5$.

Note that all we do is take the digits after the decimal point and multiply by n.

Exercises 29.4

Calculate the following.

(i) $16 \bmod 5$, (ii) $22 \bmod 4$, (iii) $-33 \bmod 22$, (iv) $7 \bmod 7$, (v) $545 \bmod 12$.

One feature of modular arithmetic is that we can often attack theorems by the method of cases, as the next theorem shows.

Theorem 29.5

If n is a square number, then $n \bmod 4$ is 0 or 1.

Proof. By definition, if n is a square, then $n = k^2$ for some integer k. As we are working mod 4 we will consider the 4 cases for $k \bmod 4$. These are 0, 1, 2, and 3.

Case $k \bmod 4 = 0$: Here $k = 4m$, for some integer m. Then, $n = k^2 = 16m^2 = 4(4m^2) = 0 \bmod 4$.

Case $k \bmod 4 = 1$: Here $k = 4m + 1$. Then, $n = k^2 = 16m^2 + 8m + 1 = 4(4m^2 + 2m) + 1 = 1 \bmod 4$.

Case $k \bmod 4 = 2$: Here $k = 4m + 2$. Then, $n = k^2 = 16m^2 + 16m + 4 = 4(4m^2 + 4m + 1) = 0 \bmod 4$.

Case $k \bmod 4 = 3$: Here $k = 4m + 3$. Then, $n = k^2 = 16m^2 + 24m + 9 = 4(4m^2 + 6m + 2) + 1 = 1 \bmod 4$.

$\square$

The next theorem allows us a different way of finding when two numbers are equal modulo n.

Theorem 29.6

We have $x = y \bmod n$ if and only if $x - y = kn$, for some $k \in \mathbb{Z}$.

Proof. Suppose $x = y \bmod n$. Then, by definition,

$$x = k_1 n + r \text{ and } y = k_2 n + r \text{ for some } k_1, k_2, r \in \mathbb{Z}.$$

Thus, $x - y = (k_1 n + r) - (k_2 n + r) = (k_1 - k_2)n$.

Now we prove the converse. If y has remainder r when divided by n, then we have $y = k_1 n + r$ for some $k_1 \in \mathbb{Z}$. So,

$$
\begin{aligned}
& x - y = kn \text{ for some } k \in \mathbb{Z} \\
\iff \quad & x = y + kn \\
\iff \quad & x = k_1 n + r + kn \\
\iff \quad & x = (k_1 + k)n + r.
\end{aligned}
$$

Thus, x has remainder r when divided by n, so $x = y \bmod n$. $\square$

Exercises 29.7

You should apply the earlier methods to the theorem and proof. In particular you should do the following.

(i)   Create examples and non-examples for the statement of the theorem.
(ii) Identify where the assumptions are made in the proof. (In the converse it is not stated explicitly where we used the assumption.)

Remark 29.8

Some mathematicians define two numbers x and y to be equal modulo n if $x - y = kn$, for some $k \in \mathbb{Z}$. The above theorem shows that this definition and ours are equivalent.

The arithmetic of mod

We now start doing some arithmetic with mod. It is easy to define addition, subtraction and multiplication of remainders to produce a theory. The concept of division is harder to handle so we ignore it.

Theorem 29.9

Suppose that $x = r \bmod n$ and $y = s \bmod n$. Then

(i) $x + y = r + s \bmod n$, *and*
(ii) $xy = rs \bmod n$.

Proof. By assumption we have $x = kn + r$ and $y = ln + s$ for some integers k and l. Then,

$$x + y = kn + r + ln + s$$
$$\implies (x + y) - (r + s) = (k + l)n.$$

Therefore, (i) follows from Theorem 29.6.
      Similarly,

$$xy = (kn + r)(ln + s)$$
$$= kln^2 + kns + lnr + rs$$
$$xy - rs = (kln + ks + lr)n,$$

so (ii) is also true.   □

Through using negatives in (i) i.e. $x - y$ is $x + (-y)$, we can also deal with subtraction.

Exercises 29.10

(i)   This is a theorem and proof so analyse it. Find examples, extreme examples, and so on.
(ii) Prove that if $x = y \bmod n$, then $kx = ky \bmod n$ for all integers k.

We can use the theorem to get quick answers to hard calculations.

Example 29.11

Find the last digit of 8^7.

Here we note that the last digit of a number is the remainder on division by 10. Thus we need to work mod 10. We know that $8^7 = 8^3 8^4$.

We have $8^2 = 64$, and $64 = 4 \bmod 10$. Thus

$$8^3 \bmod 10 = 8^2 8 \bmod 10 = (4 \times 8) \bmod 10 = 32 \bmod 10 = 2 \bmod 10.$$

Also,

$$8^4 \bmod 10 = (8^2)^2 \bmod 10 = 4^2 \bmod 10 = 16 \bmod 10 = 6 \bmod 10.$$

Thus

$$8^7 \bmod 10 = 8^3 8^4 \bmod 10 = (2 \times 6) \bmod 10 = 12 \bmod 10 = 2 \bmod 10.$$

Hence, the last digit of 8^7 is 2.

Exercise 29.12

Find the last digit of 17^{10}.

Fermat's Little Theorem

Fermat's Little Theorem[1] is a useful theorem in number theory. Like the examples above it allows investigation of very large numbers that are hard to calculate by hand. For example, we can use it to show that $2^{101} - 2$ is divisible by 101.

This time we shall see a polished version of the proof and the explanations will be given in boxes.

Theorem 29.13 (Fermat's Little Theorem)

Let p be a prime, then $x^p = x \bmod p$ for any integer x.

The usual exhortations apply: try some examples, make sure you understand what the theorem says.

Proof. We shall prove the statement for natural numbers by using mathematical induction on x. The general case is left as an exercise. Note that we can't do induction on p since $p + 1$ is only prime for $p = 2$.

Easy parts of proofs often get left as exercises. Make sure that you do them.

The initial case is $x = 0$. The statement holds as

$$0^p = 0 = 0 \bmod p.$$

---

[1] This is different to Fermat's Last Theorem, which was described earlier.

The initial part of an induction is often that easy.

Thus, let us assume that $x^p = x \bmod p$ for some $x \in \mathbb{N}$ and consider what happens for $x + 1$. We expand $(x + 1)^p$ via the Binomial Theorem.

$$(x + 1)^p = x^p + \binom{p}{1}x^{p-1} + \binom{p}{2}x^{p-2} + \cdots + \binom{p}{p-1}x + 1.$$

To use induction we require $(x+1)^p = x+1 \bmod p$, so we have taken the complicated side of the equation to see what we can do. There appears to be nothing to do other than expand it and to do that we need the Binomial Theorem.

Look on the right-hand side of the equation. You will see that we have x^p. We know by the inductive hypothesis that when we take $\bmod p$ then $x^p = x \bmod p$. We also have a 1 on the right-hand side, so we just need to get rid of all the other stuff – i.e. show that it is zero modulo p.

We produce the following which you should go through trying to see why each step is true. At each step ask 'Why?' You may need to fill in some gaps or use earlier theorems.

Taking $\bmod p$ of both sides we find

$$(x + 1)^p \bmod p = x^p + \binom{p}{1}x^{p-1} + \binom{p}{2}x^{p-2} + \cdots + \binom{p}{p-1}x + 1 \bmod p$$

$$= x + \binom{p}{1}x^{p-1} + \binom{p}{2}x^{p-2} + \cdots + \binom{p}{p-1}x + 1 \bmod p$$

$$= (x + 1) + \left(\binom{p}{1}x^{p-1} + \binom{p}{2}x^{p-2} + \cdots + \binom{p}{p-1}x\right) \bmod p$$

$$= x + 1 \bmod p.$$

The last line follows since $\binom{p}{r} = 0 \bmod p$ for $1 \le r \le p - 1$ by Exercise 24.10(x).

Does that last statement hold true? Be active! Go back and check that the exercise really does give us that.

Thus, by the Principle of Mathematical Induction, the statement is true. □

Exercise 29.14

We have proved the statement only for natural numbers. Now prove the theorem for when x is an integer. Use the trick that if $x < 0$, then $-x$ is a natural number and so the statement holds for $-x$.

Some mathematicians use the following corollary as the statement of Fermat's Little Theorem.

Corollary 29.15

Let p be a prime which does not divide the integer x, then $x^{p-1} = 1 \bmod p$.

Exercises 29.16

(i)   This statement is not equivalent to our statement. Is it weaker or stronger? Explain.

(ii)   Prove this corollary. Note that you need to use $p \nmid x$ as some point.

Another corollary of Fermat's Little Theorem is that, for a prime p, $x^p - x$ is divisible by p. This is easy to see as $x^p = x \bmod p$ is equivalent to $x^p - x = 0 \bmod p$ and this implies by Theorem 29.6 that p divides $x^p - x$.

Here are some examples.

Examples 29.17

(i) $4^5 - 4 = 1020$ is divisible by 5.

(ii) $2^7 - 2 = 126$ is divisible by 7.

(iii) $2^{100} - 1 = 1\,267\,650\,600\,228\,229\,401\,496\,703\,205\,375$ is divisible by 101. Here we see that 2 does not divide 101 and so, by Corollary 29.15, $2^{100} = 1 \bmod 101$, thus $2^{100} - 1$ is divisible by 101.

Applications of Fermat's Little Theorem

We shall now look at some more applications of Fermat's Little Theorem (FLT).

Finding remainders

Example 29.18

Calculate the remainder of 3^{293} when divided by 97.

To answer this, first notice that $293 = (3 \times 97) + 2$. We have

$$3^{293} = 3^{3 \times 97 + 2} = 3^2 \times (3^3)^{97} = 9 \times 27^{97}.$$

By FLT we know that $27^{97} = 27 \bmod 97$. Hence,

$$\begin{aligned}
3^{293} \bmod 97 &= \left(9 \times 27^{97}\right) \bmod 97 \\
&= (9 \times 27) \bmod 97 \\
&= 243 \bmod 97 \\
&= 49 \bmod 97.
\end{aligned}$$

The last reduction is worked out by hand. Note that it is a lot simpler than doing the calculation for 3^{293}. Hence, 3^{293} has remainder 49 when divided by 97.

By the way, 3^{293} is equal to 625932688882434275223255726484171540770295819045895914364335574203790567798349687214449959559620216681529421310461792605 45730559106399531123.

Primality testing

Fermat's Little Theorem may just appear as a cute little result about numbers: 'Hey, if you raise x to a prime and reduce modulo the prime, you get back x.' But what else is it good for? Let's analyse the second version of the statement from an 'if ..., then ...' point of view. We have

> If p is a prime, then $x^{p-1} = 1 \bmod p$.

If you are thinking like a mathematician, then your automatic system should have been triggered and you should have asked 'Is the converse true?' (see Chapter 16, How to read a theorem). The answer is that it is not, and I'll leave you to find a counterexample. Consider instead the contrapositive statement:

> If $x^{p-1} \neq 1 \bmod p$, then p is not a prime.

We know that this statement is true because it is the contrapositive of a true statement. But what good is it? Well, we know that primes are important and having some way to check whether a number is prime would be great. And that is what the contrapositive statement gives us.

Example 29.19

Show that 39 is not prime.

Let us take $x = 2$. Then we wish to show that $2^{38} \neq 1 \bmod 39$.

To calculate 2^{38} we rewrite 38 in some other form and use it to calculate. For example, $38 = 32 + 6 = (2^5) + 6$ or $38 = 36 + 2 = 6^2 + 2$ whatever makes $2^{38} \bmod n$ easy to calculate.

We have

$$2^{38} = 2^{36+2} \bmod 39$$
$$= (2^6)^6 2^2 \bmod 39.$$

We can calculate that $2^6 \bmod 39 = 64 \bmod 39 = 25 \bmod 39$. We have $25^2 \bmod 39 = 625 \bmod 39 = 1 \bmod 39$, and so $25^6 \bmod 39 = (25^2)^3 \bmod 39 = 1^3 \bmod 39 = 1 \bmod 39$.

From this we deduce that $2^{38} \bmod 39 = (2^6)^6 2^2 \bmod 39 = (1 \times 4) \bmod 39 = 4 \bmod 39 \neq 1 \bmod 39$.

Of course, this is a poor way to decide whether or not 39 is prime. However, it can be good in a theoretical setting or for when the suspected non-prime is large such as in the following exercise.

Exercises 29.20

(i)   Show that 66 013 is not a prime.
(ii) If x^{p-1} is 1 modulo p, then can we conclude that p prime?

Divisibility tests

As we have seen in earlier examples we can use modular arithmetic to investigate divisibility of numbers. We shall now look at other examples of this. In the following two examples we use the fact that we can write any number as a sum of multiples of powers of ten. For example,

$$374 = (3 \times 100) + (7 \times 10) + 4 = (3 \times 10^2) + (7 \times 10) + 4.$$

Example 29.21

An integer with at least two digits is divisible by 4 if and only if the number consisting of the last two digits is divisible by 4:

For an integer x we have $x = 100a + b$ where b is the number that gives the last two digits of x. For example, $x = 83\,756 = (837 \times 100) + 56$.

Then

$$
\begin{aligned}
x \bmod 4 &= 100a + b \bmod 4 \\
&= (4 \times 25)a + b \bmod 4 \\
&= 4 \times 25 \times a \bmod 4 + b \bmod 4 \\
&= b \bmod 4.
\end{aligned}
$$

Thus x is divisible by 4 if and only if b, i.e. the number formed by the last two digits, is divisible by 4.

Example 29.22

An integer is divisible by 9 if and only if the sum of its digits is divisible by 9.

We can write the integer x in decimal form: $x = \sum_{i=0}^{\infty} a_i 10^i$ where $0 \le a_i \le 9$ and a_i represents the ith digit of x.

Then,

$$
\begin{aligned}
x \bmod 9 &= \sum_i a_i 10^i \bmod 9 \\
&= \sum_i a_i 1^i \bmod 9, \ \text{since } 10 = 1 \bmod 9, \\
&= \sum_i a_i \bmod 9.
\end{aligned}
$$

Before the age of calculators this used to give a method for checking answers to products and sums. See the exercises below on 'casting out nines'.

Exercises

Exercises 29.23

(i)   Prove that an integer is divisible by 6 if and only if it is divisible by 2 and 3.

(ii)   Prove that an integer is divisible by 5 if and only if its last digit is 0 or 5.

(iii)   Prove that an integer is divisible by 3 if and only if the sum of its digits is divisible by 3.

(iv)   Prove the following.
   (a) If the difference between the sum of all digits in even positions and the sum of all digits in odd positions is zero or a multiple of 11, then the number is a multiple of 11.
   (b) No prime except 11 is a palindrome if it has an even number of digits.

(v)   In the previous exercise identify the assumptions and conclusions used. Can the hypotheses be relaxed? If so, prove the statements; if not, give counterexamples.

(vi)   Simplify the proof of Theorem 29.5 by taking $k = 4m + r$, calculating $n^2 = (4m + r)^2$ and then dealing with the cases.

(vii) For all $x \in \mathbb{Z}$ prove that $x^2 + 5x = 0 \bmod 4 \iff x^3 - 3x = 0 \bmod 4$.

(viii) Prove that $x^3 = x \bmod 6$ for all $x \in \mathbb{Z}$.

(ix)   Before the age of calculators the method of 'casting out nines' was taught as a method of checking calculations. The idea was that we could check the addition or multiplication of x and y. We add the digits of x and if we have a 9, then we ignore it, i.e. cast it out. For example, for $x = 2\,465\,981$, we ignore the 9, we can ignore the 8 and 1 as $8 + 1 = 9$, similarly we cast out the 4 and 5, thus $2\,465\,981$ reduces to $2 + 6 = 8$. In general, this resulting integer we call $s(x)$. So, in our example, $s(2\,465\,981) = 8$.
   (a) Prove that, for all integers x, $s(x)$ equals x modulo 9.
   (b) Hence, prove that $x + y = s(x) + s(y) \bmod 9$ and $xy = s(x)s(y) \bmod 9$.
   (c) We can use this to check our calculations. If we have $x + y \neq s(x) + s(y) \bmod 9$, then we have made a mistake. Does $x + y = s(x) + s(y) \bmod 9$ mean that the calculation is correct?

(x)   Suppose that n_1 and n_2 are coprime integers greater than 1. Prove that for integers a and b the simultaneous equation $ax = b \bmod n_1$ and $ax = b \bmod n_2$ has a solution $x \in \mathbb{Z}$ if and only if there is a solution to $ax = b \bmod n_1 n_2$.

(xi)   (a) Prove that if $x = y \bmod n$, then $x^k = y^k \bmod n$ for all $k \in \mathbb{N}$.
   (b) Prove that $3^{4n+1} + 2^{8n+3} = 1 \bmod 5$ for all $n \in \mathbb{N}$.
   (c) Prove that $5^{3n} + 2^{n+1} = 0 \bmod 3$ for all $n \in \mathbb{N}$.
   (d) Prove that $2^{3n} + 6 = 0 \bmod 7$ for all $n \in \mathbb{N}$.
   (e) Prove that $6^{2n} - 3 \times 6^n + 3 = 1 \bmod 5$ for all $n \in \mathbb{N}$.

(xii) Prove or disprove the following.
   (a) For all integers x and y and natural numbers n, if $xy = 0 \bmod n$, then $x = 0 \bmod n$ or $y = 0 \bmod n$.
   (b) For natural numbers $n > 3$ and integers x, if $x^2 = 4 \bmod n$, then $x = 2 \bmod n$.
   (c) There exists a natural number n such that for all integers x, if $x^2 = 4 \bmod n$, then $x = 2 \bmod n$.

(xiii) Prove that every prime greater than 3 is 1 less or 1 more than a multiple of 6.

(xiv) Prove that $(x + y)^p = x^p + y^p \bmod p$ for all integers x and y and primes p.

Summary

▶ $x = y \bmod n$ if the two integers x and y have the same remainder when we divide by a natural number n.

▶ $x = y \bmod n$ if and only if $x - y = kn$, for some $k \in \mathbb{Z}$.
▶ If $x = r \bmod n$ and $y = s \bmod n$, then
  (i) $x + y = r + s \bmod n$,
  (ii) $x - y = r - s \bmod n$, and
  (iii) $xy = rs \bmod n$.
▶ Fermat's Little Theorem: Let p be a prime, then $x^p = x \bmod p$ for any integer x.

CHAPTER 30

Injective, surjective, bijective – and a bit about infinity

Listen, there are no measurements in infinity. You humans have got such limited little minds.
The Doctor in *Doctor Who and the Masque of Mandragora*

We have seen that sets are building blocks of mathematics and have said a little about functions between sets. We shall now look more closely at functions. For functions $f : X \to Y$ we define injective, surjective and bijective functions. These definitions allow us to compare sets and in the case of bijective functions allow us to say whether one set is just a relabelling of the elements of the other.

Furthermore, using the notion of a bijection we can define two *different* types of infinite sets, those for which we can count the elements, such as $\mathbb{N}$, and those for which we can't count, such as $\mathbb{R}$. Thus we have two types of infinity!

Injective functions

Definition 30.1

*A function $f : X \to Y$ is called **injective** or **one-to-one**[1] if, for all $x_1 \in X$, $x_2 \in X$, $x_1 \neq x_2$ implies that $f(x_1) \neq f(x_2)$.*

The definition says that if I take two elements of X, then their values under f are the same if and only if the elements are the same. What we do not want is, for example, $f(3) = f(5)$. A picture of an injective function and a non-injective function is given in Figure 30.1.

Remark 30.2

In practice we do not work with the definition but with its contrapositive:

A function $f : X \to Y$ is injective if, for all $x_1 \in X$, $x_2 \in X$, $f(x_1) = f(x_2)$ implies that $x_1 = x_2$.

---

[1] The name one-to-one can be very misleading since for *any* function an element of X goes to an element of Y. My colleague Alan Slomson says that one-to-one functions would be better being called two-to-two functions since the idea is that if we take any two distinct elements of X they should go to two distinct elements of Y.

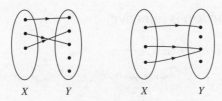

Figure 30.1 (i) Injective function (ii) Non-injective function

This is easier to work with as we can assume $f(x_1) = f(x_2)$ and proceed directly to see what this implies.

Examples 30.3

(i) The function $f : \mathbb{R} \to \mathbb{R}$ given by $f(x) = 5x - 3$ is injective.
Suppose that $f(x_1) = f(x_2)$; then we have

$$
\begin{aligned}
& f(x_1) = f(x_2) \\
\implies \quad & 5x_1 - 3 = 5x_2 - 3 \\
\implies \quad & x_1 = x_2.
\end{aligned}
$$

Thus f is injective.

(ii) The function $f : \mathbb{R} \to \mathbb{R}$ given by $f(x) = x^2$ is not injective.
(To show this we need just one counterexample.) Since $f(-1) = (-1)^2 = 1^2 = f(1)$ we see that this is not injective.

(iii) The function $f : \mathbb{N} \to \mathbb{R}$ given by $f(x) = x^2$ is injective. Suppose that $f(x_1) = f(x_2)$, then we have

$$
\begin{aligned}
& f(x_1) = f(x_2) \\
\implies \quad & x_1^2 = x_2^2 \\
\implies \quad & x_1^2 - x_2^2 = 0 \\
\implies \quad & (x_1 - x_2)(x_1 + x_2) = 0.
\end{aligned}
$$

Thus $f(x_1) = f(x_2)$ implies $x_1 = x_2$ or $x_1 + x_2 = 0$. If the former, then f is injective. If the latter, then $x_1 = -x_2$, but if $x_2 \in \mathbb{N}$, then $x_1 \notin \mathbb{N}$. Thus only the former holds.

Comparing this example with the previous shows the importance of the domain of definition for the map. We should not just look at the formula defining f.

(iv) Let $f : \mathbb{R}\backslash\{1\} \to \mathbb{R}$ be given by $f(x) = x/(x - 1)$. Then f is injective.
Suppose that $f(x_1) = f(x_2)$; then we have

$$
\begin{aligned}
& f(x_1) = f(x_2) \\
\implies \quad & \frac{x_1}{x_1 - 1} = \frac{x_2}{x_2 - 1} \\
\implies \quad & x_1(x_2 - 1) = x_2(x_1 - 1) \\
\implies \quad & x_1 x_2 - x_1 = x_2 x_1 - x_2 \\
\implies \quad & -x_1 = -x_2 \\
\implies \quad & x_1 = x_2.
\end{aligned}
$$

How to show that a function is injective

To show that a function is injective is simple:

> Suppose that $f(x_1) = f(x_2)$. Then show by direct implications that $x_1 = x_2$.

Surjective functions

Definition 30.4

*A function $f : X \to Y$ is called **surjective** or **onto** if for all $y \in Y$ there exists $x \in X$ such that $f(x) = y$.*

A schematic diagram of a surjective function and a non-surjective function is given in Figure 30.2.

From our knowledge of quantifiers we know that if someone gives us a y from Y, then we have to find an x in X so that $f(x) = y$ is true.

Examples 30.5

(i)   The map $f : \mathbb{R} \to \mathbb{R}$ given by $f(x) = x^3$ is surjective.

Suppose that $y \in \mathbb{R}$. We require that $f(x) = y$, i.e. $x^3 = y$. We can just take $x = y^{1/3}$. Let's write this properly:

Suppose that $y \in \mathbb{R}$. Then let $x = y^{1/3}$. We have $f(x) = f(y^{1/3}) = (y^{1/3})^3 = y$. Thus f is surjective.

(ii)   The function $f : \mathbb{R} \to \mathbb{R}$ given by $f(x) = 5x - 3$ is surjective.

Since the definition of surjectivity involves an existential quantifier, then for all $y \in \mathbb{R}$ we have to *find* an x so that $f(x) = y$. We are trying to solve an equation where y is given. Thus we want x so that $5x - 3 = y$. Obviously we can rearrange this so that x is the subject of the equation: $x = (y+3)/5$. Let us write this for public consumption:

The function f is surjective. Suppose $y \in \mathbb{R}$, then let $x = (y + 3)/5$. We have

$$
\begin{aligned}
f(x) &= f\left(\frac{y+3}{5}\right) \\
&= 5\left(\frac{y+3}{5}\right) - 3 \\
&= y + 3 - 3 \\
&= y.
\end{aligned}
$$

(iii)   The map $f : \mathbb{R} \to \mathbb{R}$ given by $f(x) = x^2$ is not surjective. Let $y = -1$, then there is no x such that $f(x) = -1$ since this would require x such that $x^2 = -1$.

(iv)   The map $f : \mathbb{R} \to \mathbb{R}^+ = \{z \in \mathbb{R} \mid z \geq 0\}$ given by $f(x) = x^2$ is surjective.

Again, given y in $\{z \in \mathbb{R} \mid z \geq 0\}$ we have to find x such that $x = y^2$. Now since y is non-negative we can take $x = \sqrt{y}$.

This example and the previous show how important the codomain is when dealing with surjectivity.

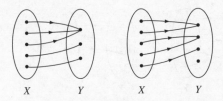

Figure 30.2 (i) Surjective function (ii) Non-surjective function

How to show that a function is surjective

To show that a function is surjective is simple:

> We *suppose* that y is in the codomain and then *show* that there exists an x in the source so that $f(x) = y$. This usually involves solving this equation. We then use the solution of this equation early in our written proof.

Remark 30.6

A common error is to 'prove' a function is surjective by showing that if $x \in X$, then there exists a $y \in Y$ such that $f(x) = y$. That is, we take an element in X and find one in Y. However, this is the definition of a function! Remember that for surjectivity we have to take elements in Y and find in X what maps to them.

Composition of functions

Now we investigate what happens when we apply one function after another.

Definition 30.7

*Suppose that $f : X \to Y$ and $g : Y \to Z$ are two functions. Then the **composition** of f and g, denoted $g \circ f$, is the map from X to Z defined by $(g \circ f)(x) = g(f(x))$. That is, we apply f and then apply g.*

Examples 30.8

(i)   Let $f : \mathbb{N} \to \mathbb{R}$ be given by $f(x) = 1/x$. Let $g : \mathbb{R} \to \mathbb{R}$ be given by $g(x) = x^2$. Then $(g \circ f)(x) = g(f(x)) = g(1/x) = (1/x)^2 = 1/x^2$.

(ii) Let $f : \mathbb{N} \to \mathbb{N}$ be given by $f(x) = x + 1$. Let $g : \mathbb{N} \to \mathbb{N}$ be given by $g(x) = x^2$. Then $(g \circ f) = g(f(x)) = g(x + 1) = (x + 1)^2 = x^2 + 2x + 1$.

Remarks 30.9

(i)   If $g \circ f$ is defined, then this does not mean that $f \circ g$ is defined since the domain of f may be different to the target of g.

(ii) If $g \circ f$ and $f \circ g$ are defined, then in general $g \circ f \neq f \circ g$.

Exercise 30.10

Give examples to show that the above two statements are correct.

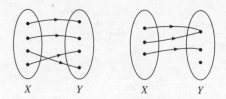

Figure 30.3 (i) Bijective function (ii) Non-bijective function

Bijective functions

When given two properties about an object, say injective and surjective, then the true mathematician asks what do we get when we combine them.

Definition 30.11

*A function $f : X \to Y$ is called **bijective** (or **a bijection**, or a **one-to-one corres-pondence**) if f is injective and surjective.*

The key to understanding the definition lies in the word correspondence. The idea is that the elements of X and Y correspond precisely to one another.

Examples 30.12

(i)   The map $f : \mathbb{R} \to \mathbb{R}$ given by $f(x) = x^3$ is bijective. From previous examples we know that it is both injective and surjective.

(ii) Again by previous examples we know that $f : \mathbb{R} \to \mathbb{R}$ given by $f(x) = 5x - 3$ is injective and surjective, hence it is bijective.

(iii) The map $f : \mathbb{N} \to \mathbb{N}$ given by $f(x) = x^2$ is injective but not surjective (and so not bijective).

Exercises 30.13

(i)   Prove that the map $f : \mathbb{R} \to \mathbb{R}$ defined by $f(x) = ax + b$ is bijective for all $a, b \in \mathbb{R}$ with $a \neq 0$.

(ii) Give an example of a map that is surjective but not injective.

How to show that a function is bijective

To show that a function is bijective is simple:

Show it is injective and surjective.

Inverse functions

An interesting property of a bijective map $f : X \to Y$ is that there exists a special map from Y to X called the inverse.

Definition 30.14

*The **inverse function** (or **inverse map**) of a bijective function $f : X \to Y$ is the map $f^{-1} : Y \to X$ given by: $f^{-1}(y)$ is the unique element x of X such that $f(x) = y$.*

Note that in the definition we use surjectivity and injectivity: for an element $y \in Y$ we know that there is an element x such that $f(x) = y$ as the map is surjective, and we know that this x is unique by injectivity.

Examples 30.15

(i) Consider the example $f : \mathbb{R} \to \mathbb{R}$ given by $f(x) = x^3$. The inverse function is $f^{-1}(y) = y^{1/3}$.

(ii) The inverse to $f : \mathbb{R} \to \mathbb{R}$ given by $f(x) = 5x - 3$ is $f^{-1}(y) = (y + 3)/5$.

As you can see in these two examples the inverse is given by making x the subject of the equation $f(x) = y$.

Exercises 30.16

(i) Show that the sine and cosine functions when considered as maps from $\mathbb{R}$ to $\mathbb{R}$ are neither injective nor surjective.

(ii) Consider the sine function as being restricted to the set $[-\pi, \pi]$ with codomain $[-1, 1]$. That is, $\sin : [-\pi, \pi] \to [-1, 1]$. Show that this is a bijection.

(iii) Can you find a domain and codomain upon which cosine is bijective?

Theorem 30.17

Suppose that $f : X \to Y$ is bijective. Then

(i) $f^{-1} \circ f$ *is the identity map on X, that is, $(f^{-1} \circ f)(x) = x$ for all $x \in X$, and*

(ii) $f \circ f^{-1}$ *is the identity map on Y, that is, $(f \circ f^{-1})(y) = y$ for all $y \in Y$.*

Exercise 30.18

Prove this theorem.

Warnings!

One of the problems with notation in mathematics is that the standard notation is sometimes inconsistent.

Problems with trigonometrical notation

Since the sine function restricted to the set $[-\pi, \pi]$, $\sin : [-\pi, \pi] \to [-1, 1]$, is a bijective function it has an inverse. Using our convention that the inverse for f is f^{-1} then the inverse is $\sin^{-1}$. Thus the inverse value of x is $\sin^{-1}(x)$. (This is also known as arcsin.) However, it is conventional to use $\sin^2(x)$ to denote $(\sin(x))^2$ and, more generally, $\sin^n(x)$ to denote $(\sin(x))^n$. What then happens if $n = -1$? Here we have $\sin^{-1}(x)$ denoting the inverse value of x and the expression $1/\sin(x)$. Obviously we have to be careful in such situations to avoid confusing the two.

Problems with preimage set

Further problems occur in that f^{-1} is also used as notation when dealing with sets.

Definition 30.19

*Let $f : X \to Y$ be a function and S be a subset of Y. Then the **preimage** of S is the set $f^{-1}(S)$ given by*

$$f^{-1}(S) = \{x \in X \mid f(x) = y \text{ and } y \in S\}.$$

If S is a single element of Y, say y, then we can write $f^{-1}(y)$.

Note that $f^{-1}(S)$ is a set. It is a subset of X.

Examples 30.20

(i)   Let f be function $f : \mathbb{R} \to \mathbb{R}$ given by $f(x) = x^2$.

    (a) If $S = \{5\}$, then $f^{-1}(S) = \{-\sqrt{5}, \sqrt{5}\}$.
    (b) If $S = \{-5\}$, then $f^{-1}(S) = \emptyset$.
    (c) If $S = \{0\}$, then $f^{-1}(S) = \{0\}$.
    (d) If $S = [1, 2]$, then $f^{-1}(S) = [-\sqrt{2}, -1] \cup [1, \sqrt{2}]$.
    (e) If $S = [0, 2]$, then $f^{-1}(S) = [-\sqrt{2}, \sqrt{2}]$.

(ii) Let f be the sine function $\sin : \mathbb{R} \to [-1, 1]$. Then $f^{-1}(0) = \{n\pi \mid n \in \mathbb{Z}\}$.

Notice that the preimage of a set is defined for all functions; it does not require that f is a bijection.

Problems arise since f^{-1} can denote a function or it could be used in denoting a set. It will usually be clear from context what is required. The point is that you have to be observant and be careful.

Types of infinity – countable and uncountable

Now that we have defined bijectivity of a function we can define different types of infinity. The basic idea is that there are different types of infinity: infinities that we can count and infinities that we can't count. The natural numbers (the counting numbers) are our basic example of an infinite set that we can count. The set of real numbers is an example of an infinite set where we can't count the elements.

Definition 30.21

*We say that an infinite set X has **countably many elements** if there exists a bijection between X and $\mathbb{N}$. We also say that X is **countable** or **countably infinite**.*

*If X is an infinite set which is not countable, then we say X is **uncountable** or **uncountably infinite**.*

Examples 30.22

(i)   The natural numbers are obviously countable as the identity map id : $\mathbb{N} \to \mathbb{N}$ given by $\text{id}(n) = n$ is a bijection.

(ii) The set of even numbers is countable. The map $f : \text{Even} \to \mathbb{N}$ given by $f(x) = x/2$ is a bijection. Similarly, the set of odd numbers is countable. You can give the bijection!

Notice that the set of even numbers is a proper subset of $\mathbb{N}$ but we can find a bijection between it and $\mathbb{N}$. This sort of bizarre counter-intuitive behaviour is common when dealing with infinity.

Exercises 30.23

(i)   Show that the set of square numbers, $\{1, 4, 9, 16, 25, \dots\}$, is countable.

(ii) Show that the set $\{1, 1/2, 1/3, 1/4, \dots, 1/n, \dots\}$ is countable.

We should start thinking like mathematicians and find other sets that are countable and find those that are not. For example, is $\mathbb{Z}$ countable? Since we have $\mathbb{N} \subset \mathbb{Z}$ our intuition may be to say that there can be no bijection between the two. In fact, I have heard the (mistaken) argument given that '$\mathbb{N}$ is infinite in one direction, i.e. positive, while $\mathbb{Z}$ is infinite in two directions, i.e. positive and negative, and so they can't be of equal size.'

   In fact what we do to make the required bijection clear is count the elements of $\mathbb{Z}$ in a special way. We list them as $0, 1, -1, 2, -2, 3, -3, 4, -4, \dots$.

Exercises 30.24

(i)   Explicitly give a bijection that gives this counting and show it is a bijection. (Hint: use a definition by cases depending on n even or odd.)

(ii) Prove or disprove: If X is countably infinite and $Y \subseteq X$, then Y is countably infinite.

The rationals are countable

Proving that the set of positive rational numbers $\mathbb{Q}$ is countable is a little bit harder. To begin with we use the idea of using a special way of ordering fractions to show that the positive rational numbers are countable.

   Imagine that we write all possibilities for a/b in a grid as in Figure 30.4. Certainly all rational numbers must be represented on this grid. However, some are represented twice, for example, $2/2$ is the same as $1/1$.

   The idea is that we give an order to all these fractions. This order is described by the arrows. Start at $1/1$, then go down to $2/1$. Then back up to the top line. As you can see, the arrows cause us to snake around the grid. Notice that we manage to visit every number in the grid.

   Now, we obviously have a bijection between these pairs and $\mathbb{N}$. This is not what we want though as some fractions are the same, i.e. we have repetitions. No matter. Just throw away any repetition as we reach it. This does give us a bijection between the naturals and the positive rationals. Thus, both can be counted. They are, in effect, the same size of infinity, though one might naively think that there are more positive rationals than naturals.

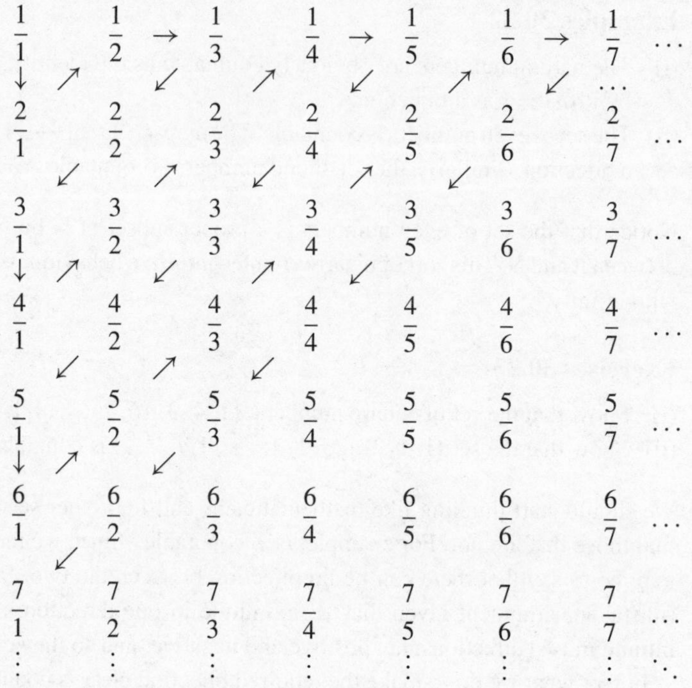

$$\frac{1}{1} \quad \frac{1}{2} \rightarrow \frac{1}{3} \quad \frac{1}{4} \rightarrow \frac{1}{5} \quad \frac{1}{6} \rightarrow \frac{1}{7} \cdots$$

$$\frac{2}{1} \quad \frac{2}{2} \quad \frac{2}{3} \quad \frac{2}{4} \quad \frac{2}{5} \quad \frac{2}{6} \quad \frac{2}{7} \cdots$$

$$\frac{3}{1} \quad \frac{3}{2} \quad \frac{3}{3} \quad \frac{3}{4} \quad \frac{3}{5} \quad \frac{3}{6} \quad \frac{3}{7} \cdots$$

$$\frac{4}{1} \quad \frac{4}{2} \quad \frac{4}{3} \quad \frac{4}{4} \quad \frac{4}{5} \quad \frac{4}{6} \quad \frac{4}{7} \cdots$$

$$\frac{5}{1} \quad \frac{5}{2} \quad \frac{5}{3} \quad \frac{5}{4} \quad \frac{5}{5} \quad \frac{5}{6} \quad \frac{5}{7} \cdots$$

$$\frac{6}{1} \quad \frac{6}{2} \quad \frac{6}{3} \quad \frac{6}{4} \quad \frac{6}{5} \quad \frac{6}{6} \quad \frac{6}{7} \cdots$$

$$\frac{7}{1} \quad \frac{7}{2} \quad \frac{7}{3} \quad \frac{7}{4} \quad \frac{7}{5} \quad \frac{7}{6} \quad \frac{7}{7} \cdots$$

Figure 30.4 Arranging the rationals in a grid

Exercise 30.25

(i) Prove that the set of rationals, $\mathbb{Q}$, is countable.

(ii) Prove that $\mathbb{N} \times \mathbb{N}$ is countable.

The reals are uncountable

We need an example of an uncountable set. To make things easy we take the set of real numbers from 0 to 1. This is obviously infinite, but is it countable? In fact, it is not.

We do a proof by contradiction. Suppose that a bijection between $\mathbb{N}$ and $[0, 1]$ existed. Then, we could list the elements of $[0, 1]$. Let us do so using their decimal expansions so that x can be written as $0.x_1 x_2 x_3 x_4 x_5 \ldots$ where x_i is a digit from 0 to 9.

Suppose that we can list the elements of $[0, 1]$

$$0 \,.\, a_1 \quad a_2 \quad a_3 \quad a_4 \quad a_5 \quad a_6 \ldots$$
$$0 \,.\, b_1 \quad b_2 \quad b_3 \quad b_4 \quad b_5 \quad b_6 \ldots$$
$$0 \,.\, c_1 \quad c_2 \quad c_3 \quad c_4 \quad c_5 \quad c_6 \ldots$$
$$0 \,.\, d_1 \quad d_2 \quad d_3 \quad d_4 \quad d_5 \quad d_6 \ldots$$
$$0 \,.\, e_1 \quad e_2 \quad e_3 \quad e_4 \quad e_5 \quad e_6 \ldots$$
$$0 \,.\, f_1 \quad f_2 \quad f_3 \quad f_4 \quad f_5 \quad f_6 \ldots$$
$$\vdots$$

Now, let us construct a new rational number from $[0, 1]$ that is *not* in this list. Call it x. Let the first digit of x after the decimal point be any digit *not* equal to a_1 or 9. Let the second digit be anything *not* equal to b_2 or 9. Let the third digit be anything *not* equal to c_3 or 9. And so on. Always take the kth digit of x to be anything other than 9 or the kth digit of the kth element of the list.

Since x is of the form $0.x_1x_2x_3x_4x_5\ldots$ we know it is between 0 and 1. So it must be on the list. It can't be the first number on the list as its first digit is different to a_1. It can't be the second number on the list as its second digit is not equal to the second digit of b. And so on. It can't be the kth element on the list as its kth digit is different to the kth digit of the kth element of the list. Thus the decimal expansion of x is not on the list. Furthermore it is not equal to anything on the list in the case that we have 9 as a repeating decimal, e.g. $0.24999999\ldots$ is $0.2500000\ldots$ just as $0.99999\cdots = 1$, since no digit in x is equal to 9.

This contradicts that we can list the elements of $[0, 1]$.

Exercise 30.26

Use the above result to show that show $\mathbb{R}$ is not countable.

Remarks 30.27

(i) From the proof that $[0, 1]$ is uncountable we note that to prove a set is uncountable it is a good idea to assume it is countable and aim for a contradiction.

(ii) The countable set $\{1, 1/2, 1/3, 1/4, \ldots, 1/n, \ldots\}$ is a proper subset of the uncountable $[0, 1]$ and so we are justified in saying that some infinities are bigger than others.

Exercises

Exercises 30.28

(i) Determine whether the following functions are injective, surjective or bijective. For any bijections write down the inverse.

(a) $f : \mathbb{R} \to \mathbb{R}$ given by $f(x) = x^2 + x + 1$.

(b) $f : \mathbb{N} \to \mathbb{N}$ given by $f(x) = x^2 + x + 1$.

(c) $f : \mathbb{R} \to \mathbb{R}$ given by $f(x) = x^3$.

(d) $f : \mathbb{R} \to \mathbb{R}$ given by $f(x) = \begin{cases} 1/x & \text{for } x \neq 0 \\ 0 & \text{for } x = 0. \end{cases}$

(e) $f : \mathbb{R} \to \mathbb{R}$ given by $f(x) = \dfrac{3x - 4}{x^2 + 5}$.

(f) $f : \mathbb{R}\backslash\{-2\} \to \mathbb{R}\backslash\{5\}$ given by $f(x) = \dfrac{5x}{x + 2}$.

(g) $f : \mathbb{R} \to \mathbb{R}^+$ given by $f(x) = |x - 3| + 3$.

(h) $f : \mathbb{R}^+ \to \mathbb{R}^+$ given by $f(x) = 2^x$.

(ii) For each map $f : \mathbb{R} \to \mathbb{R}$ above that is not a bijection can you find a and b such that the map $g : [a, b] \to f([a, b])$, defined by $g(x) = f(x)$ for all $x \in [a, b]$, is a bijection? Determine the inverse of each g.

(iii)   Describe the compositions $f \circ g, g \circ f, f \circ f$ and $g \circ g$, when they can be defined, for
    (a) $f : \mathbb{R} \to \mathbb{R}$ defined by $f(x) = x^2 - 1$, and $g : \mathbb{R} \to \mathbb{R}$ defined by $g(x) = x + 2$;
    (b) $f : \mathbb{R} \to (0, 1)$ defined by $f(x) = 1/(x^2 + 1)$, and $g : (0, 1) \to (0, 1)$ defined
        by $g(x) = 1 - x$;

(iv)   Which of the following are true? Give a proof for true statements and a counterexample for false ones.
    (a)   The composition of two injective functions is injective.
    (b)   The composition of two surjective functions is surjective.
    (c)   The composition of two bijective functions is bijective.
    (d)   The composition of an injective function and a surjective function is bijective.

(v)   Let $f : \mathbb{R} \to \mathbb{R}$ and $g : \mathbb{R} \to \mathbb{R}$ be injective. Give counterexamples to
    (a)   the function $h : \mathbb{R} \to \mathbb{R}$ given by $h(x) = f(x)g(x)$ is injective, and
    (b)   the function $h : \mathbb{R} \to \mathbb{R}$ given by $h(x) = f(x) + g(x)$ is injective.
    What happens for surjective functions?

(vi)   Theorem 30.17 was basically 'f bijective $\implies$ certain conditions' so we can ask whether the converse true. In this case we can't write the conditions without first defining what f^{-1} is. Thus one possible way of making an 'if and only if' statement is:
        'Suppose that $f : X \to Y$ is a function. Then f is bijective if and only if there exists a function $g : Y \to X$ such that
    (a) $g \circ f$ is the identity map on X, that is, $(g \circ f)(x) = x$ for all $x \in X$, and
    (b) $f \circ g$ is the identity map on Y, that is, $(f \circ g)(y) = y$ for all $y \in Y$.
    In this case $g = f^{-1}$, the inverse function for f.'
        Prove or disprove this statement.

(vii)   Consider the functions $\log_e x : (0, \infty) \to \mathbb{R}$ and $e^x : \mathbb{R} \to (0, \infty)$. Are these injective, surjective or bijective?

(viii)   Suppose that A and B are infinite sets. Prove or give a counterexample to the following statements.
    (a)   If A and B are both countable, then $A \times B$ is countable.
    (b)   If A is countable and $B \subseteq A$, then B is countable.
    (c)   If A and B are countable, then $A \cup B$ is countable.
    (d)   If A and B are countable, then $A \cap B$ is countable.
    (e)   If $B \subseteq A$, then $A \backslash B$ is finite.
    (f)   If $B \subseteq A$, then $A \backslash B$ is infinite.
    (g)   The set $\mathbb{R} \backslash \mathbb{Q}$ is uncountable. That is, the set of irrational numbers is uncountable.

Summary

▶   A function $f : X \to Y$ is injective if and only if, for all $x_1 \in X, x_2 \in X, f(x_1) = f(x_2)$ implies that $x_1 = x_2$.

▶ f is surjective if for all $y \in Y$ there exists $x \in X$ such that $f(x) = y$.

▶   The composition of $f : X \to Y$ and $g : Y \to Z$, denoted $g \circ f$, is the map from X to Z defined by $(g \circ f)(x) = g(f(x))$.

▶ f is called bijective if f is injective and surjective.

▶ f^{-1} has two meanings! It is the inverse function of f or is used in defining preimage sets.

▶ An infinite set X is countable if there exists a bijection between X and $\mathbb{N}$.

▶ An infinite set X is uncountable if it is not countable.

▶ $\mathbb{N}$ and $\mathbb{Q}$ are countable.

▶ $[0, 1]$ and $\mathbb{R}$ are uncountable.

CHAPTER 31

Equivalence relations

All animals are equal, but some animals are more equal than others.
George Orwell, *Animal Farm*, 1946

In this chapter we see how mathematics becomes abstracted, that is, we identify some property that we like from an example and then make a general definition that covers this example but also allows us to investigate a far wider group of objects. In Chapter 29, Modular arithmetic, we took the idea that two numbers were equivalent if they had the same remainder on division by n. Now we are going to abstract the whole notion of what it means to be equivalent. We will call this an equivalence relation. As an example, in a standard deck of playing cards we can say that two cards are equivalent if they have the same suit, or we can say that they are equivalent if they have the same value.

The concept of equivalence relation is hard to grasp as it is unlike any other met in lower-level mathematics. A particular sticking point is the notion of equivalence class. This is where we gather together all the equivalent elements and treat them as a single entity. For example, if we declare two cards equivalent if they have the same suit, then the equivalence class of the seven of Spades consists of all the Spades. If we take the equivalence relation to be that cards are equivalent if they have the same value, then the equivalence class of the seven of Spades is all the sevens: {seven of Spades, seven of Hearts, seven of Clubs, seven of Diamonds}.

Treating this class as a new object to play with is unlike anything met before, but once grasped leads to new and exciting mathematics, for example quotient groups in group theory, or quotient spaces in topology.

Relations

It is easy to give an intuitive idea of a relation. We can give many examples.

Examples 31.1

(i)   The number x is related to y if $x = y$.
(ii)   The integer x is related to the integer y if $x = y \bmod n$.
(iii) The integer x is related to the integer y if $x \mid y$.

(iv) The person A is related to B if A is a sibling of B.

(v)   The real number x is related to the real number y if $x < y$.

(vi) The real number x is related to the real number y if $xy = 0$.

(vii) The city X is related to city Y if there is a flight from X to Y on a Sunday.

There are many relations we can define but how do we define the notion of relation in a mathematical way? We are obviously taking a set, say the set of integers, natural numbers, people, or cities and doing something with it. Let's call this set X. Well, we take two elements so we are really working with $X \times X$. We can see this in the abstract definition of a relation.

Definition 31.2

*Let X be a set and let R be a subset of the product $X \times X$. We say x is **related** to y by the **relation** R and write $x \sim y$ if $(x, y) \in R$.*

Example 31.3

Suppose that we have the set $X = \mathbb{Z}$ and take the relation $\sim$ to be 'The integer x is related to the integer y if $x = y \bmod n$.' So, if $n = 5$ for example, then $(2, 7) \in R$ and $(101, 21) \in R$ but $(3, 4) \notin R$ and $(101, 20) \notin R$.

Note that, in the definition, R is the relation not the $\sim$. However, in practice we drop the reference to R and talk about $\sim$ as a relation on X. In practice most mathematicians rarely think about a relation as being a subset!

Equivalence relations

We can now look at a certain type of relation that is very important in mathematics.

Definition 31.4

*A relation $\sim$ on X is called an **equivalence relation** if the following three conditions hold.*

(i)   *Reflexive condition: $x \sim x$ for all $x \in X$.*

(ii) *Symmetric condition: $x \sim y \implies y \sim x$ for all $x, y \in X$.*

(iii) *Transitive : $x \sim y$ and $y \sim z$ implies that $x \sim z$ for all $x, y, z \in X$.*

Thus, given a relation all we have to do is check these conditions. Why are these good conditions to work with? Over many years mathematicians discovered that these three were simple and very common. Furthermore, as we shall see, they allow us to group elements together into disjoint sets.

Example 31.5

For the natural number n define the relation $\sim$ on $\mathbb{Z}$ by

$$x \sim y \iff x = y \bmod n.$$

We check the conditions:

(i)   Reflexive: $x \sim x \iff x = x \bmod n$. As $x = x \bmod n$ is always true we see that $\sim$ is reflexive.

(ii) Symmetric: $x \sim y$ implies that $x = y \bmod n$, and we know that this implies that $y = x \bmod n$, which just says $y \sim x$. So $\sim$ is symmetric.

(iii) Transitive: Suppose that $x \sim y$ and $y \sim z$, then we have $x = y \bmod n$ and $y = z \bmod n$. From the definition of mod , i.e. all the numbers have the same remainder on division by n, we deduce that $x = z \bmod n$, i.e. $x \sim z$. That is, $\sim$ is transitive.

All three conditions are satisfied, therefore $\sim$ is an equivalence relation.

Examples 31.6

(i)   The relation $>$ on the set of real numbers is not an equivalence relation since it is not reflexive: $x > x$ does not hold for any x.

(ii) The relation $\geq$ on the set of real numbers is not an equivalence relation since it is not symmetric: for example, $4 \geq 3$ is true but $3 \geq 4$ is not.

(iii) The relation 'A is a grandfather of B' is not transitive. This is because if X is a grandfather of Y and Y is a grandfather of Z, then X is not necessarily a grandfather of Z.

Exercise 31.7

For the following relations decide whether they are reflexive, symmetric or transitive. State which are equivalence relations.

(i)   The real number x is related to the real number y if $x = y$. (Note that it's a definition so implicitly it is an 'if and only if' not just an 'if'.)

(ii)   The real number x is related to the real number y if $x < y$.

(iii) The real number x is related to the real number y if $x \leq y$.

(iv)   The real number x is related to the real number y if $xy = 0$.

(v) $x \sim y$ if x is a sibling of y.

(vi)   The set A is related to B if $A \subset B$.

(vii) For $x, y \in \mathbb{R}$, x is related to y if $x - y$ is an integer.

(viii) Let $X = \mathbb{R}$. Then for all $x, y \in X$, $x \sim y$ if $|x - y| \leq 1$.

Exercise 31.8

For each condition, reflexive, symmetric, and transitive, give an example of a relation for which the condition holds but the other two do not.

Next, for each condition give an example of a relation for which the condition does not hold but the other two do.

A subtlety

If we play around with the three conditions of reflexive, symmetric and transitive, then it may appear that can we deduce reflexivity from the other two. To see this, suppose for a relation that symmetry and transitivity hold. If $x \sim y$, then by symmetry $y \sim x$. But by transitivity (using $z = x$) we can deduce that $x \sim x$.

Thus it looks as if the reflexive condition is redundant. But it is not! Look at the start of the argument. It says 'If $x \sim y...$'. That is a big if! There is nothing to guarantee that x is related to a distinct other element! (And that includes itself!) Therefore, the argument is false.

The previous exercise asked for a relation which was symmetric and transitive but not reflexive. Such a relation is an example of this behaviour.

Equivalence classes

The notion of equivalence class is very important in mathematics. It allows us to group equivalent elements together, as we shall see in the next section.

Definition 31.9

*The **equivalence class** of x under the equivalence relation $\sim$, denoted $[x]$, is the set*

$$[x] = \{y \in X \mid y \sim x\}.$$

Examples 31.10

(i) We have seen in Example 31.5 that we can use modular arithmetic to put an equivalence relation on $\mathbb{Z}$. The equivalence classes are

$$[0] = \{\dots, -3n, -2n, -n, 0, n, 2n, 3n, \dots\}$$
$$[1] = \{\dots, -3n + 1, -2n + 1, -n + 1, 1, n + 1, 2n + 1, 3n + 1, \dots\}$$
$$[2] = \{\dots, -3n + 2, -2n + 2, -n + 2, 2, n + 2, 2n + 2, 3n + 2, \dots\}$$
$$[3] = \{\dots, -3n + 3, -2n + 3, -n + 3, 3, n + 3, 2n + 3, 3n + 3, \dots\}$$
$$\vdots$$
$$[n - 1] = \{\dots, -3n + (n - 1), -2n + (n - 1), -n + (n - 1),$$
$$n - 1, n + (n - 1), 2n + (n - 1), 3n + (n - 1), \dots\}$$
$$= \{\dots, -2n - 1, -n - 1, -1, n - 1, 2n - 1, 3n - 1, 4n - 1, \dots\}$$
$$[n] = [0].$$

(ii) Let X be the set of cards in a standard deck of playing cards. Let $\sim$ be defined by $x \sim y$ if x and y are in the same suit. Then there are four distinct equivalence classes: the Spades, the Hearts, the Clubs and Diamonds.

Handy tip 31.11

Given an equivalence relation, ask 'What do the equivalence classes look like?'

Exercise 31.12

Let $X = \mathbb{R}^2$ be the plane. Consider the two relations:

(i) Point $p_1 \in \mathbb{R}^2$ is related to $p_2 \in \mathbb{R}^2$ if their distance to the origin is the same.

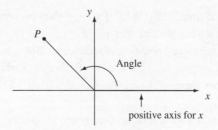

Figure 31.1 The angle for a point

(ii) For a point p we define its angle to be the angle between the line from p to the origin and the line made by the positive part of the x axis. See Figure 31.1. Point $p_1 \in \mathbb{R}^2$ is related to $p_2 \in \mathbb{R}^2$ if their angles are the same.

Show that these are equivalence relations. What are the equivalence classes? Draw pictures to represent the equivalence classes.

Partitions

We now come to the point of equivalence relations. In this section we shall see that an equivalence relation divides up a set so that each part is distinct and contains elements which to all intents and purposes are the same. For example, farmers do not put all their animals in the same field. They divide up the land so that the cows are in one part, the chickens in one part, the pigs in another and so on. That way the animals are easier to manage. Obviously within a category, for example, the cows will be different, they will have their own names and personalities, but to all intents and purposes, they are the same when it comes to milking!

Using an equivalence relation is a very powerful way of organizing a set. It divides it up into pieces that are more manageable. Obviously in any mathematics problem one has to select the right equivalence relation – for example, a farmer would not separate the animals based on the number of legs they have.

For an equivalence relation on X, the equivalence classes divide up X so that the union of the equivalence classes is X. The classes for x and y either are disjoint or are equal. Thus the set breaks up into distinct pieces. See Figure 31.2.

Let's put this in a mathematical form.

Theorem 31.13

Suppose that $\sim$ is an equivalence relation defined on X. Then

(i) $\bigcup_{x \in X}[x] = X$ *(the notation in the equation means take the union of the sets $[x]$ for all $x \in X$),*

(ii) $x \sim y$ *if and only if $[x] = [y]$, and*

(iii) $x \nsim y$ *if and only if $[x] \cap [y] = \emptyset$.*

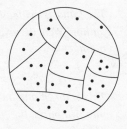

Figure 31.2 A partition for an equivalence relation

Proof. (i)   Let $x \in X$. Since $x \sim x$, then $x \in [x]$, so $x \in \bigcup_{x \in X}[x]$. Therefore, $X \subseteq \bigcup_{x \in X}[x]$. The reverse inclusion, that $\bigcup_{x \in X}[x] \subseteq X$, is obvious.

(ii) [$\Rightarrow$] Suppose that $x \sim y$. If $z \in [x]$, then $z \sim x$ and so by the transitive property $z \sim y$, thus $z \in [y]$. This shows $[x] \subseteq [y]$. But if $x \sim y$, then by the symmetry property $y \sim x$ and by reasoning similar to the above we find $[y] \subseteq [x]$. Therefore, $[x] = [y]$.

(ii) [$\Leftarrow$] As $x \sim x$ then $x \in [x] = [y]$ so $x \sim y$.

(iii) [$\Rightarrow$] Suppose that $x \not\sim y$ and assume for a contradiction that $[x] \cap [y] \neq \emptyset$. As $[x] \cap [y] \neq \emptyset$ there exists $z \in X$ such that $z \in [x]$ and $z \in [y]$, in other words $z \sim x$ and $z \sim y$. By the transitivity property $x \sim y$. This is a contradiction.

(iii) [$\Leftarrow$] Suppose that $[x] \cap [y] = \emptyset$ and for a contradiction that $x \sim y$. As $x \sim y$, then $x \in [x]$ and $x \in [y]$, thus $x \in [x] \cap [y]$. This is a contradiction. $\square$

Remark 31.14

The theorem shows that the sets $[x]$ and $[y]$ are non-empty and that they are either equal or disjoint.

Let's go through that proof but this time analysing it.

Proof. *To begin with let us note that our overall assumption is that $\sim$ is an equivalence relation, i.e. satisfies the three conditions: reflexive, symmetric and transitive. When analysing the proof we need to check that all these were used. If they were not, then we could weaken the assumption to, for example, $\sim$ is reflexive and symmetric.*
    Analysis of (i):

Let $x \in X$.	*Trying to show that two sets are equal so will work at level of elements.*
Since $x \sim x$,	*Reflexive property used! So one part of the assumption has been used.*
then $x \in [x]$,	*Definition of equivalence class used.*
so $x \in \bigcup_{x \in X}[x]$. Therefore, $X \subseteq \bigcup_{x \in X}[x]$.	*Effectively we are trying to show that two sets A and B are equal, so the problem has been divided into $A \subseteq B$ and $B \subseteq A$; see Chapter 20. We have just shown the former by working at the level of elements. Need to do the latter!*

The reverse inclusion, that
$\bigcup_{x \in X}[x] \subseteq X$, is obvious.

The latter is evidently obvious to the author! We will have to check it at some point.

Analysis of (ii):

[$\Rightarrow$]

This is an 'if and only if' so gets divided into two parts, the 'only if' comes first. That is, $x \sim y \implies [x] = [y]$. The assumption is $x \sim y$; we will need to check that it is used.

Suppose that $x \sim y$.

Looks like the assumption is being used. But it is not. The author is just stating that the assumption will be used.

If $z \in [x]$, then $z \sim x$ and so by the transitive property $z \sim y$,

Definition of equivalence class used. Ah, here's where we use the assumption that $x \sim y$. And we use the transitive condition too!

thus $z \in [y]$.
This shows $[x] \subseteq [y]$.

Definition of equivalence class used again. Aha, using the '$A \subseteq B$ and $B \subseteq A$ gives $A = B$' trick again. Got one inclusion, need the other!

But if $x \sim y$, then by the symmetry property $y \sim x$
and by reasoning similar to the above we find $[y] \subseteq [x]$.
Therefore, $[x] = [y]$.

Used symmetry condition, so have used all three conditions for equivalence relation. We'll need to check that this is correct. Seems ok. Got the conclusion we want from the assumption that $x \sim y$.

(ii) [$\Leftarrow$]
As $x \sim x$
then $x \in [x] = [y]$

Now need to do the 'if' part of (ii). Uses reflexivity again. Uses two ideas at once. It uses the definition of equivalence class, i.e. $x \in [x]$, and then uses the assumption in the $\Leftarrow$ statement, i.e. $[x] = [y]$.

so $x \sim y$.

Got the conclusion we want by using $x \in [y] \implies x \sim y$.

Analysis of (iii):

(iii) [$\Rightarrow$]

Again, (iii) is an 'if and only if' so it gets broken into two pieces.

Suppose that $x \nsim y$ and assume for a contradiction that $[x] \cap [y] \neq \emptyset$.
As $[x] \cap [y] \neq \emptyset$ there exists $z \in X$ such that $z \in [x]$ and $z \in [y]$,

Using proof by contradiction so assuming that conclusion for 'only if' part is false. Reasonable since we are assuming it for a contradiction.

in other words $z \sim x$ and $z \sim y$. By the transitivity property $x \sim y$. This is a contradiction.

Used again!
One of our assumptions was that $x \not\sim y$.

(iii) [⇐] Suppose that $[x] \cap [y] = \emptyset$ and for a contradiction that $x \sim y$.
As $x \sim y$, then $x \in [x]$ and $x \in [y]$,

Another proof by contradiction!

Using definition of equivalence relation.

thus $x \in [x] \cap [y]$. This is a contradiction.

Well, we assumed that the intersection was empty.

$\square$

Our analysis of this proof shows that we have used the assumption that $\sim$ is an equivalence relation. So that is good. It does not mean that if we weaken the statement it will be false. This idea will be discussed in Chapter 33. The point is that *our* proof used the three conditions. How do we know that there is not someone out there who has a proof which uses only one or two of the conditions? What we need to show is that if we drop any one of the three conditions for an equivalence relation, then there exist counterexamples to conclusions from the theorem.

Exercise 31.15

Find these counterexamples!

The theorem was that if we have an equivalence relation, then we have a way of dividing up the set into pieces so that the whole set is made up of pieces and these pieces are disjoint. Of course, we can think like mathematicians and ask if the converse is true.

Therefore we ask 'If we have some way of dividing the set, then can we get an equivalence relation?' Yes, we can. First we need to define what we mean by dividing a set.

First, let us note that for a countable collection of sets we can use the notation X_1, X_2, X_3, ... to denote the sets. However, a collection may be indexed by an uncountable set. Say we let $X_a = [-a, a]$ where $a \in \mathbb{R}$. This collection of sets is uncountable because $\mathbb{R}$ is uncountable. To cater for both we can index a collection by taking the sets X_λ where $\lambda \in I$ and I is some set. For example, I could be $\mathbb{N}$ or $\mathbb{R}$ or $\mathbb{R} \times \mathbb{N}$, etc.

Definition 31.16

Let X be a set. Let X_λ be a collection of subsets where $\lambda \in I$ for some set I and such that

(i) $\bigcup_{\lambda \in I} X_\lambda = X$,
(ii) $X_\lambda = X_\mu$ *if and only if* $\lambda = \mu$,
(iii) $X_\lambda \cap X_\mu = \emptyset$ *if and only if* $\lambda \neq \mu$.

*The collection of sets is called a **partition** of X.*

Example 31.17

For an equivalence relation the set of equivalence classes is a partition. This is from Theorem 31.13 by setting $X_x = [x]$ and $I = X$.

Now let's get a converse to Theorem 31.13. That is, given a partition we define an equivalence relation.

Exercise 31.18

Suppose X is partitioned by $\{X_\lambda\}_{\lambda \in I}$. Define the relation $\sim$ by $x \sim y$ if $x \in X_\mu$ for some μ implies that $y \in X_\mu$. Prove that this is an equivalence relation.

What we have shown is that putting an equivalence relation on a set is equivalent to giving a partition and vice versa. This is how I think about equivalence relations. We parcel up the set into pieces – the equivalence classes – and do mathematics with these classes.

The whole point of equivalence relations is that we can partition the set and concentrate only on the equivalence classes; we bundle up the elements into a single package.

Modular arithmetic

We saw in Example 31.5 that $x \sim y \iff x = y \bmod n$ is an equivalence relation and saw in Examples 31.10 that the equivalence classes are $[0], [1], [2], \ldots, [n-1]$. This set of equivalence classes is denoted by $\mathbb{Z}_n$.

We can now define arithmetic on the classes by the following: Let

$$[x] + [y] = [x + y]$$
$$[x] \cdot [y] = [xy].$$

Note that what we are doing is defining arithmetic on the set of equivalence classes, not on the set of integers. We are saying that we multiply the equivalence classes $[x]$ and $[y]$ by defining the product to be the equivalence class of xy. See Figure 31.3(i).

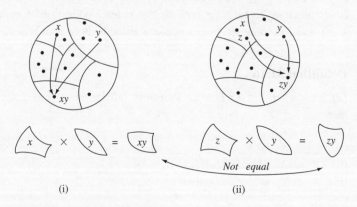

(i)   (ii)

Figure 31.3 Multiplying classes

Examples 31.19

(i) In $\mathbb{Z}_7$ we have $[3] \cdot [5] = [3 \times 5] = [15] = [1]$.

(ii) In $\mathbb{Z}_8$ we have have $[6] + [7] = [6 + 7] = [13] = [5]$.

Now we come to a deep, deep idea in mathematics. We want to multiply the *classes* but our definition actually depended on the *elements* in X, not the classes. For example, what we want is that if $[x] = [z]$, then $[z] \cdot [y] = [x] \cdot [y]$. We do not want the resulting class to depend on the specific element we took in the class. For example, it should not depend on whether we took $[1]$ or $[6]$ when working mod 5. It is certainly possible that we may have that $x \sim z$ but $xy \nsim zy$, see Figure 31.3(ii).

Theorem 31.20

If $[x] = [u]$ and $[y] = [v]$, then $[u] \cdot [v] = [x] \cdot [y]$.

Proof. Suppose that $x \sim u$ and $y \sim v$. Then we must show that $xy \sim uv$.

We have $x \sim u \iff x = k_1 n + u$ for some $k_1 \in \mathbb{Z}$ by definition, and similarly $y \sim v \iff y = k_2 n + v$ for some $k_2 \in \mathbb{Z}$. Then

$$\begin{aligned}
xy &= (k_1 n + u)(k_2 n + v) \\
&= k_1 k_2 n^2 + k_1 n v + u k_2 n + uv \\
&= (k_1 k_2 n + k_1 v + k_2 u) n + uv.
\end{aligned}$$

Thus $xy \sim uv$. $\qquad\square$

In other words, the definition of multiplication does not depend on which elements of the classes we use.

Definition 31.21

*An element used to represent an equivalence class is called a **representative** of the class.*

Definition 31.22

*Suppose we have an operation defined on classes using the elements of the classes. If the definition does not depend on the representatives we say that the operation is **well-defined**.*

Example 31.23

Multiplication of classes is well-defined by the above theorem.

Exercise 31.24

Define addition of classes by $[x] + [y] = [x + y]$. Show that this is well-defined.

Exercises

Exercises 31.25

(i) For the following relations on X determine whether they are reflexive, symmetric or transitive. State whether they are equivalence relations or not and if they are describe their equivalence classes.

(a) Let $X = \mathbb{Z}$ and $x \sim y \iff x - y$ is even.

(b) Let $X = \mathbb{Z}$ and $x \sim y \iff x - y$ is odd.

(c) Let $X = \mathbb{R}$ and $x \sim y \iff xy = 0$.

(d) Let $X = \mathbb{R}$ and $x \sim y \iff xy \neq 0$.

(e) Let $X = \mathbb{Z} \times \mathbb{Z}$ and $(x_1, y_1) \sim (x_2, y_2) \iff x_1 = x_2$.

(f) Let $X = \mathbb{Z} \times \mathbb{Z}$ and $(x_1, y_1) \sim (x_2, y_2) \iff x_1 y_1 = x_2 y_2$.

(g) Let $X = \mathbb{R}$ and $x \sim y \iff x - y \in \mathbb{Q}$.

(h) Let $X = \mathbb{Z}$ and $x \sim y \iff 5 | (x + 3y)$.

(i) Let $X = \mathbb{Z}$ and $x \sim y \iff \text{Int}(x) = \text{Int}(y)$ where $\text{Int}(x)$ is the largest integer such that $\text{Int}(x) \leq x$.

(j) Let $X = \mathbb{R} \times \mathbb{R}$ and $(a, b) \sim (c, d) \iff (a - c)(b - d) = 0$.

(k) Let $X = \mathbb{R} \times \mathbb{R}$ and $(a, b) \sim (c, d) \iff (a - d)(b - c) = 0$.

(ii) Determine which of the following classes are equal in (a) $\mathbb{Z}_6$ and (b) $\mathbb{Z}_9$:

$$[-114], [-3], [-1], [0], [1], [3], [6], [8], [9], [12], [17], [15], [27], [45].$$

(iii) For $m \in \mathbb{N}$ define a function $f : \mathbb{Z}_m \to \mathbb{Z}_m$ by $f([x]) = [x + x^2]$. Show that it is well-defined.

(iv) Determine whether the following are well-defined.

(a) The map $f : \mathbb{Z}_8 \to \mathbb{Z}_6$ given by $f([x]) = [2x]$.

(b) The map $f : \mathbb{Z}_3 \to \mathbb{Z}_9$ given by $f([x]) = [x^2]$.

(c) The map $f : \mathbb{Z}_m \to \mathbb{Z}_m$ given by $f([x]) = [ax + b]$, where $a, b \in \mathbb{Z}_m$.

Summary

▶ A relation on a set X is a subset R of $X \times X$. The element $x \in X$ is related to $y \in X$, denoted $x \sim y$, if $(x, y) \in R$.

▶ An equivalence relation is a relation that is

   (i)   reflexive, $x \sim x$,

   (ii) symmetric, $x \sim y \implies y \sim x$,

   (iii) transitive, $x \sim y$ and $y \sim x \implies x \sim z$.

▶ The equivalence class of x is $[x] = \{y \in X \mid y \sim x\}$.

▶ An equivalence relation partitions a set into disjoint pieces: the equivalence classes.

▶ We can define operations on classes using representatives but must check that the operation is well-defined.

PART VI

Closing remarks

CHAPTER 32

Putting it all together

*Nothing has such power to broaden the mind as the ability
to investigate systematically and truly all that comes under thy observation in life.*
Marcus Aurelius, *Meditations*

In the preceding five parts of this book there is a lot to digest and each part focuses on a different way of thinking like a mathematician. Putting them all together and applying them effectively takes time and practice. It is not easy to think like a mathematician; there is no set of rules to follow, though many techniques have been given. Doing mathematics is certainly more of an art than a science.

Therefore, in this chapter I will pick out some interesting points from earlier chapters and see how we can integrate the methods.

Getting started

The biggest hurdle with any mathematics problem is getting started. I set weekly exercises for my students and when a student says 'I am stuck on question X' my standard reply is 'Well, what have you done so far?' Their reply is often 'Nothing. I can't get started. I don't even know where to start.'

The advice in many Study Aid type books is to look for previous worked examples. However, my students don't need to have this pointed out to them, they have probably already looked in their notes for a similar problem. And in fact this study book advice may be the *cause* of their difficulties.

Too often, students beginning high-level mathematics look for an example or a theorem from which they can deduce a statement as a simple corollary. Unfortunately, mathematical thinking is not like that.

With this in mind, let's look at some techniques for getting started. (We'll leave aside trivial problems such as not understanding all the words in the question!)

Play with examples

Rarely can you say *nothing* about a particular problem. There may be some example or some deduction you can make that relates to it. So the first technique for getting started is to play with examples. This often provides some illumination. Somehow just playing with some examples allows us a deeper understanding of the problem. I am not sure why this works. Maybe it forces us to engage with the question without actually solving it.

Consider Theorem 20.2, which you may recall was 'Let A and B be finite sets. Then $|A \cup B| = |A| + |B| - |A \cap B|$.'

The proof for this theorem is given earlier but let us suppose it wasn't. Playing with some examples very quickly shows why the equality holds and then the method of proof quickly becomes obvious.

If a statement involves infinite sets, then see what happens for particular examples of infinite sets like $\mathbb{N}$ and $\mathbb{R}$. Doing this sort of exercise gets the mathematical part of your brain going and working on the problem.

Break it down – take small bites

Another reason that some students find it hard to get started: they are going for the 'big bite' approach. They expect that there is a single algorithm/calculation/formula/theorem that will answer the question. They do not see that they need to take little bites and often need to use a collection of algorithms, calculations, formulas and theorems.

Thus a guiding principle should be to break a problem down into smaller problems. For example, given 'Prove $A \Longleftrightarrow B$' do not look for a theorem with this conclusion or try to create a list of the form $A \Longleftrightarrow C$, $C \Longleftrightarrow D$, etc., in the hope that you reach B. Instead prove $A \Longrightarrow B$ and then $B \Longrightarrow A$. We have seen this a number of times. The proofs of Theorem 29.6 and Theorem 31.13(iii) are examples of this.

The same goes for showing that $A = B$ for two sets A and B. Here our hope is that there is a chain of equalities $A = C = D = \cdots = Z = B$, which can happen; see Example 20.11.

The bite-sized method is to show $A \subseteq B$ and $B \subseteq A$. This can be seen in the proofs of Theorem 31.13 parts (i) and (ii). And how do we show that $A \subseteq B$? We go down to the level of the elements, we take $x \in A$ and show that $x \in B$; see, for example, Theorem 20.9.

A problem that is broken down like this is easier to solve, even though it appears that we have doubled the number of problems! One advantage of this method is that proving one part gives a feeling of accomplishment, so that if you are asked what have you done so far, you can say 'Well, I can show $A \Longrightarrow B$, but not the other way round.'

Change the problem

The third method to get started is to change the problem. What we do is specialize and generalize. These two methods are so important that they are given their own chapter, Chapter 33.

For the moment a good example of changing the problem is the general version of the Division Lemma on page 199. One assumption was that y was a non-zero integer. However, the bulk of the proof involved proving a different statement, the one where y is a *positive* integer. Once that part of the proof was done a separate, and shorter, argument dealt with the negative y case. This method can be used again and again. If the statement says integers it may be easier to use the natural numbers. More generally, we can restrict to a smaller class of objects and investigate those. The point is that, by changing the question, we at least can do something.

Getting to a higher level

Getting started is of course very important. But how do you go beyond merely answering questions and get to truly thinking like a mathematician? Let's see some techniques.

Reverse the question – construct your own examples

A real mathematician invents their own examples.

Here, I don't only mean finding examples of definitions and theorems as suggested in earlier chapters, useful as that is. An example of what I mean is given in calculus. Finding the maxima and minima of functions is a basic exercise in calculus: you are given a function and have to find its critical points.

Now the technique for creating examples is: Reverse that question. Suppose that you are given that the maxima and minima are at certain points. Can you construct a function which has these maxima and minima?

You will learn lots about maxima and minima from doing this. Obviously one could just guess a function and work with that. But a higher level of understanding is to be gained from creating an example with some pre-defined restrictions, e.g. a maximum at point $x = 2$ with value $f(x) = 5$ and minima at $x = -2$ and $x = 7$ with $f(-2) = 20$ and $f(7) = -3$.

For another example of this reverse-the-question method, consider conics from elementary geometry. The standard question is 'Here is a conic, calculate its centre, vertices, eccentricity' and so on. Reverse this question. Given conditions involving the centre, vertices and eccentricity create a conic with these properties.

Ask 'What happens if . . .?'

Good mathematicians like to ask 'What happens if . . .?' For example, what happens if I drop that assumption? We have seen above that this can help solve problems. It also allows us to explore the limits of a subject. We can find out why definitions and theorems are the way they are.

As another example, mathematical objects are often sets with some extra conditions. At a very simple level we can say that a finite set is one with a finite number of elements but there are much more complicated examples such as groups. (A group is a set with a way of multiplying elements of the set. The multiplication has to satisfy certain properties.)

Now, for sets A and B we can take their product $A \times B$. We can ask, if A and B have a certain property, then does $A \times B$? For example, suppose that A and B are finite sets.

Is $A \times B$ a finite set? In this case, yes it is. If A and B are infinite, then is $A \times B$ infinite? If A and B are groups, then is $A \times B$ a group? And so on.

The idea is that we are always asking questions to create new concepts.

The importance of reflection – see the web

Although it appears to be a linear subject, with one idea built on top of another, mathematics also contains a web of interconnection of ideas and topics. After learning some topic your reflection should include thinking about how the new work fits in this web. Don't just finish a topic and say 'I've done that now.'

Also, you can analyse the overall structure of the work and ask if it could have been presented in a different order. This gives a different perspective and is a useful exercise to aid understanding. Similarly, look in different books for different points of view. Are theorems and definitions different? Are the proofs, more, or less, rigorous?

Exercises

Exercises 32.1

(i)   Let F_n be the nth Fibonacci number. My friend tells me that

$$F_1 + F_2 + F_2 + \cdots + F_n = F_{n+2},$$

but he can't remember if it is really right.

Is he right? If not, then can you rescue his statement by making a similar statement? Either way, prove any statement you think is correct.

(ii) Let n be a postive integer and let $d_k, d_{k-1}, \ldots, d_1, d_0$ be its digits. Further, let f be the function defined by $f(n) = d_k^3 + d_{k-1}^3 + \cdots + d_0^3$, that is, the function which maps a positive integer n to the sum of the cubes of its digits. Given an initial n one may repeatedly apply the map f to it to generate a sequence of integers, a_m.

We say n is a **dead-end integer** if there exists m_0 such that $a_m = a_{m_0}$ for all $m \geq m_0$. If $n = 19$, for example, then

$$a_1 = 19, \ a_2 = 1^3 + 9^3 = 730, \ a_3 = 7^3 + 3^3 + 0^3 = 370,$$
$$a_4 = 3^3 + 7^3 + 0^3 = 370, \ldots$$

So $n = 19$ is a dead-end integer. Let us denote the set of dead-end integers by $\mathcal{D}$.

Any **fixed point** of the function f (i.e. n such that $f(n) = n$) is a dead-end integer. Let $\mathcal{F}$ denote the set of fixed points.

(a) Investigate the the sets $\mathcal{D}$ and $\mathcal{F}$.

(b) Are all positive integers dead-end integers?

(c) If not, can you say anything about the sequences generated by those which aren't?

(d) Is $\mathcal{D}$ bounded? Is $\mathcal{F}$ bounded?

(e) Can you determine $\mathcal{F}$ completely?

This exercise and the following one are taken from a course at the University of Leeds.

(iii) A student on a year abroad buys a number of presents and wants to send them home. The goods are already wrapped up as individual packets. By a miracle, there is an integer n such that there is precisely one packet of depth a, width b, and length c, for each set of integers (a, b, c) with $1 \leq a \leq b \leq c \leq n$.

It is cheaper (and mathematically more interesting) to send the packets wrapped up together in larger parcels. To avoid breakages, the parcels must have no empty spaces. In other words, a parcel of size $p \times q \times r$ will have volume pqr that is equal to the sum of the volumes of the packets it contains. Assuming that it does not matter how large the parcels are, how many parcels does the student need?

(iv) Can you create a mathematical method for calculating the particular day of the week given the date? For example, for 20 September 1968 and for 8 December 1978 it should calculate Friday. Hint: Think mod.

Summary

► To get started:
   (i)   Play with examples.
   (ii) Break it down – take small bites.
   (iii) Change the problem.
► To get to a higher level:
   (i)   Reverse the question – construct your own examples.
   (ii) Ask 'What happens if …?'
   (iii) Reflect and see the web of ideas.

CHAPTER 33

Generalization
and specialization

All generalizations are misleading.

Anon.

To some extent we have been generalizing and specializing throughout the book. Discussing it in detail has been delayed until now as one needs to see examples before really getting to grips with these ideas.

Generalization

Weakening the hypotheses

Given a theorem a question you should ask is 'Can I weaken the hypothesis and still get the same conclusion?' The objective of mathematics is to use as few hypotheses as possible to get as strong a conclusion as possible. That is, use as little as possible to say as much as possible.

The statement given by weakening the assumptions is called a **generalization**.

Let us consider the simple theorem on even numbers:

'If x and y are even natural numbers, then $x + y$ is even.'

We can weaken the hypothesis to the statement

'If x and y are even integers, then $x + y$ is even.'

This statement is true.

We could have weakened the hypothesis in a different direction by dropping the requirement that the numbers are even:

'If x and y are natural numbers, then $x + y$ is even.'

However, note that this statement is false; take $x = 2$ and $y = 3$ for instance. Thus weakening (or, as in this case, losing) an assumption in a theorem can lead to a false statement.

We have already seen an example where weakening a hypothesis can lead to a new theorem. The Division Lemma (page 196) was originally stated for a natural number y and then this was weakened to an integer y to get the general version of the Division Lemma (page 199).

Thus, if a theorem has an assumption that x is a natural number, then generalize to x is an integer. In fact, the proof may not be complicated. We prove the statement for natural numbers and use the nice trick we applied in the general version of the Division Lemma: for negative x we let $x' = -x$. Since x' is positive the statement is true for x'. We use this knowledge to prove the statement for x. (And we need to remember to do the $x = 0$ case!)

Changing the conclusion

If we weaken a hypothesis in statement P and get a false statement, then also changing the conclusion may give a true statement Q. If the statement P can be deduced from this Q, then we also call Q a generalization.

We have seen this in Pythagoras' Theorem: Let T be a right-angled triangle with sides of length a, b and hypotenuse c. Then $c^2 = a^2 + b^2$.

We can weaken this by dropping the right-angled condition. Doing this we can't define a hypotenuse so we produce the statement: Let T be a triangle with sides of length a, b and c. Then $c^2 = a^2 + b^2$.

This is false as almost any random example will show. However, if we change the conclusion, then we can get a true statement, the Cosine Rule: Let T be a triangle with sides of length a, b and c. Then $c^2 = a^2 + b^2 - 2ab \cos C$, where C is the angle opposite the side of length c.

Note that we can deduce Pythagoras' Theorem from the Cosine Rule. We take $C = 90°$ as an assumption and the conclusion is a special case of the Cosine Rule conclusion, i.e. $\cos 90° = 0$ so $c^2 = a^2 + b^2$.

Therefore, the Cosine Rule is a generalization of Pythagoras' Theorem. Note that the assumptions of the Cosine Rule refer to all triangles, whereas the assumptions of Pythagoras' Theorem refer to right-angled triangles.

Generalizations that are false can be interesting

Attempting to generalize can lead to interesting mathematics. We know that there are many integer solutions to $x^2 + y^2 = z^2$. But what if we change the 2 to a more general natural number n?

We get 'There exist integer solutions to $x^n + y^n = z^n$ for $n \geq 2$.' This is false. In actual fact it is not just the case that there exists some $n \geq 3$ such that there are no positive integer solutions to $x^n + y^n = z^n$, but in fact, there are no positive integer solutions for *all* $n \geq 3$. This latter statement is Fermat's Last Theorem, which we met on page 101. The attempt to prove this statement covered 350 years and led to huge advances in mathematics through number theory, algebraic geometry and so on.

Specialization

Specialization is the reverse of generalization. We make the assumptions stronger. For example, if have a statement involving integers we can restrict to the natural numbers. Similarly, Pythagoras' Theorem is a specialization of the Cosine Rule.

We can use specialization to rescue statements from being false. The statement 'If $a < b$, then $a^2 < b^2$' is not true. (Take $a = -5$ and $b = 2$, for example.) We can strengthen the assumptions by including the conditions $a > 0$ and $b > 0$. Thus we specialize to 'Let $a, b > 0$. If $a < b$, then $a^2 < b^2$.' This statement *is* true.

The main application of specialization is in solving problems. It can be easier to solve a problem for a special case and use this to gain insight. For example, the more general Division Lemma is in effect proved by specializing to the case of natural numbers. Once this is done we can generalize to the case of integers.

Without loss of generality

In a number of proofs we used the phrase 'without loss of generality'. In some texts this is sometimes abbreviated to the cryptic wlog.

We use the phrase when we make an assumption so that the new statement *looks* like a specialization but in fact it is not – the same level of generality remains. For example, if we had a statement that began 'For any two distinct integers x and y we have ...', then without loss of generality we could assume that $x < y$. If it wasn't, then we could just relabel.

Exercises

Exercises 33.1

(i)   What is the best possible assumption about n so that $\sqrt{n}$ is irrational? That is, complete the statement '$\sqrt{n}$ is irrational if and only if n is of the form ...' Justify your answer.

(ii) Consider the exercise of showing $\sqrt[7]{7!} < \sqrt[8]{8!}$ from Exercise 5.3(iv). How far can you generalize this? Generalize part (b) of the same exercise.

(iii) In a 3×3 grid arrange the numbers 1 to 9 so that the sum along any row, the sum of any column or the sum of one of the two diagonals is constant. You can use each number only once. Such a square is called a **magic square**. As an example consider

4	9	2
3	5	7
8	1	6

Any row, any column and the two diagonals add up to 15. Can you find other examples of magic squares?

Can you find a 4×4 magic square using the numbers 1 to 16? What about $n \times n$ squares using 1 to n^2?

What happens if we generalize the 3×3 case by relaxing the condition that the numbers are from 1 to 9 and instead take any integers? For the 3×3 case can the constant that the rows etc. sum to be any given integer? What about the $n \times n$ case?

(iv) Take a 4×4 grid of 16 squares. Show that you can place six crosses in the grid so that there exists an even number of crosses in each row and column.

This is sometimes referred to as the Milk Crate Problem as the original problem was stated in terms of placing milk bottles into a milk crate.

Can you generalize this problem to an $n \times n$ grid where $n \geq 3$?

(v)   Find the last digit of $3^{122}5^{600}112^{450}$. What about the penultimate digit? How far can you generalize a result about the last digit of the product of three numbers?

(vi) Generalize Exercise 26.7(v) on $ax^2 + bx + c = 0$, where $a, b, c \in \mathbb{Q}$ to a polynomial of higher degree. Can we prove one solution is rational if and only if another solution is rational?

(vii) In order to rationalize its stamp printing operation, the Post Office has decided to issue postage stamps in only two denominations, a pence and b pence, where a and b are positive integers. The disadvantage of this proposal is that (unless either a or b is 1), not every possible postage price can be made up from the stamps. For example, if $a = 5$ and $b = 8$, the post office could charge 13 pence or 16 pence, but not 17 pence. Given a particular choice of a and b, let us refer to a number as 'good' if it is a viable postage price and 'bad' if it is not. In the example above, 13 and 16 are good, while 17 is bad.

- Discover as much as you can about how the set of bad numbers depends on the choice of a and b.
- Start by experimenting with some specific choices of a and b which are not too big, e.g. $a = 5$, $b = 8$. Can you see any patterns?
- It's a good idea to be systematic when experimenting. Try fixing the value of a, $a = 5$ say, and considering each of the cases $b = 2, 3, \ldots, 10$ in turn. Assuming $a = 5$, can you prove anything about the set of bad numbers for general b? What about $a = 6$, general b?
- What can you prove for the case of general a and b?
- Can you think of a precise mathematical definition of a 'good' number? If so, can you think of a geometrical interpretation of the definition?

This problem was also taken from a University of Leeds course.

Summary

▶ If we weaken assumptions, for example drop assumptions, then we get a generalization.

▶ If we strengthen assumptions, then we get a specialization.

▶ Without loss of generality means to make what looks like a specialization, but in fact the same level of generality is maintained.

CHAPTER 34

True understanding

In mathematics you don't understand things. You just get used to them.
John von Neumann (1903–1957)

How do you know when you really understand something in mathematics? This is very hard to answer. Often one can have the feeling of understanding and yet in attempting exercises and problems one's lack of understanding soon becomes obvious. In this chapter we will list ways of demonstrating understanding.

Understanding definitions

You understand a definition if you

- can state it precisely,
- can state it in your own words,
- can give concrete examples of it, including trivial and non-trivial examples,
- can give non-examples of the definition,
- can recognize it in different and unfamiliar situations,
- know theorems in which it can be used,
- know why it can be used in those theorems,
- know why this particular definition is made,
- know other similar definitions of the same word and know the differences between them.

The last item on the list occurs because different mathematicians use different definitions and this has important consequences for theorems. By adding in an extra hypothesis to a definition many theorems become easier to prove.

Understanding theorems

You understand a theorem if you

- can state it precisely and in your own words,
- can give concrete examples of its use,

- understand its proof,
- can apply it in new and unfamiliar situations,
- can give a counterexample to statements given by weakening hypotheses,
- know its inverse and converse,
- can see some consequences from it (corollaries for example),
- can encapsulate it in one sentence, e.g. this gives me a method to calculate distance,
- see where it fits in the theory, is it an end in itself or a theorem used on the path to a greater theorem,
- know whether it refers to a small or large class of objects.

Understanding proofs

You understand a proof if you

- can state it precisely and in your own words,
- know where the assumptions are used,
- know the structure, i.e. how it breaks apart and which techniques (direct, contradiction, etc.) are used,
- can see every step as simple rather than as a miracle,
- can use the ideas in the proofs in your own proofs of other statements,
- can fill in any gaps,
- know how rigorous the proof is,
- can summarize it, i.e. leave out the details but keep key points,
- know where problems occur when hypotheses are dropped.

Understanding a major topic

Usually mathematics is grouped into different topics, e.g. calculus, differential equations, combinatorics, etc.

You understand a topic if you

- can see how it all fits together,
- can change definitions slightly to produce a different theory,
- can encapsulate it in one sentence,
- can give a concrete example which exhibits many of the features of the theory,
- can see connections, similarities and differences between this topic and others,
- can move effortlessly between an intuitive grasp and technical details in an argument,
- know what the key definition or theorem is,
- know why it is interesting and useful,
- know what is the bare minimum needed to make the theory work,
- know which ideas get used and used again in the theory,
- can explain it without notes,
- can explain it to someone else.

Exercises

Exercises 34.1

(i) Look through books, lecture notes and web pages. Judge how well you know the definitions, theorems and proofs there. Try using the above as a checklist.

(ii) Which subject in mathematics is the one that you are least confident about or understand the least? Go back to it and apply the ideas from this book!

Summary

▶ Definitions: Know examples and non-examples and which theorems the definition is used in.

▶ Theorems: Know examples and non-examples to which it applies, know what consequences it has.

▶ Proofs: Know where the assumptions are used, know how a particular proof is structured.

▶ Major topic: Know the key examples.

CHAPTER 35

The biggest secret

Do or do not. There is no try.
Yoda in *The Empire Strikes Back*

Gurus . . . make fortunes from motivational courses that are both amazing and sinister,
but which boil down to an age-old and obvious adage: just get on with it.
It's about do or don't do.
Derren Brown, *Tricks of the Mind*, 2006

In this book I have collected together many different ideas and techniques that I use time and time again as a mathematician. Owing to pressures of space, and the fact that they would make the book harder to digest, I omitted quite a few techniques. Subjects such as parity and degrees of freedom are useful in checking an answer, and topics such as tautologies and circular arguments are important in logic, but I have left them for you to discover.[1]

Like may self-help books there is much to be taken in at once, and multiple rereadings may be necessary. Even then, from such books, we often take away only one or two nuggets. Given that much has been said and much has been left unsaid, how can I sum everything up? What must you do if you really want to think like a mathematician?

The key practical advice I would give to any aspiring mathematician is in two sentences.

- Write mathematics correctly.
- Create your own examples.

These techniques are surprisingly effective. The reasons for the first are discussed in Chapter 3. Basically it boils down to: if you can't describe it properly, then the chances are that you don't understand it properly. And if you can't describe it properly, then you know to look into it and try to understand. It helps make explicit any weaknesses in your understanding.

The second piece of advice has also been dealt with in earlier chapters. Maybe this is the real secret of thinking like a mathematician – creating your own problems and examples. Trying to create a decent problem can be hard. However, you can use very simple techniques like reversing the question. If you create these problems with your

---

[1] Did you expect me to do everything for you?

friends, then you can exchange them (the problems, not the friends) and get even more practice. You can also set a competition: see who can set the hardest – yet manageable – problem.

To create your own examples actually requires a large stock of good, extreme and trivial examples, so you will need to collect them and have them at your fingertips. One way of doing this is to imagine what you would say when someone wakes you up in the middle of the night and says 'Quick it's an emergency, give me a good example of an X.'

Of course, you should feel free to decide for yourself what the most important techniques are for becoming a good mathematician: whether it is 'Choose the most complicated side of an equation and reduce' (page 143) or always to try the contrapositive (Chapter 26).

There is one last bit of advice to be given. The one big secret that separates out the mathematicians from the non-mathematicians. It is *attitude*. It takes time to explore mathematics. Often it seems more important just to plough on with the exercises that need to be handed in tomorrow. However, the material in this book saves time in the long run. It identifies what you know and what you don't know, what you need to work on, etc. If you have the attitude that understanding is crucial – not superficial understanding but the understanding that comes from a deep attack on a problem – and if you are always looking beyond what you have been given, then magically you will understand what you have been given.

Whether you use the techniques in the book is up to you. You can choose to use them or choose to ignore them. As the quotes say at the start of this chapter. Do or don't do. The choice is yours.

Happy mathematical thinking!

Summary

► Just get on with it.

Greek alphabet

In the pronunciation guide the letter i should be pronounced as in big. The letter y should be pronounced as in the word my.

Name	Upper	Lower	Pronunciation
Alpha	A	α	al-fa
Beta	B	β	bee-tah
Gamma	Γ	γ	gam-ah
Delta	Δ	δ	del-tah
Epsilon	E	ϵ or ε	ep-sigh-lon / ep-sil-on
Zeta	Z	ζ	zee-tah
Eta	H	η	ee-tah
Theta	Θ	θ	thee-tah
Iota	I	ι	y-oh-tah
Kappa	K	κ	cap-ah
Lambda	Λ	λ	lam-dah
Mu	M	μ	mew
Nu	N	ν	new
Xi	Ξ	ξ	ksy
Omicron	O	o	oh-mi-kron
Pi	Π	π	py
Rho	P	ρ	row (as in propelling a boat)
Sigma	Σ	σ	sig-mah
Tau	T	τ	taw
Upsilon	Υ	υ	up-sigh-lon or up-sil-on
Phi	Φ	ϕ or φ	fy
Chi	X	χ	ky
Psi	Ψ	ψ	psy
Omega	Ω	ω	oh-meg-ah

Commonly used symbols and notation

e	base of natural logarithms
i	square root of -1
∞	infinity
$\forall$	for all
$\exists$	there exists
$\square$	end of proof marker
∇	nabla
$\aleph$	aleph
$\emptyset$	empty set
$\sum$	sum
$\prod$	product
$\in$	is an element of / is in
$\subseteq$	is a subset of
$\subset$	is a subset of (but is not equal to)
$\cap$	intersection
$\cup$	union
$\Rightarrow$	implies that (often incorrectly used, see page 37)
$\Leftrightarrow$	equivalent to (also known as 'if and only if')
$\prime$	prime
$\mapsto$	maps to
$\twoheadrightarrow$	surjection
$\hookrightarrow$	injection
$\propto$	proportional to
$\equiv$	equivalent/congruent to
$\approx$	approximately equal
$\perp$	perpendicular to
$\neg$	negation
$\sim$	tilde
$\hat{}$	hat
$\therefore$	therefore
$\because$	because

$\mathbb{N}$ natural numbers, defined as the set $\{1, 2, 3, 4, \dots\}$ (but see the footnote on page 4)

$\mathbb{Z}$ integers, defined as $\{\dots, -3, -2, -1, 0, 1, 2, 3, \dots\}$

$\mathbb{Q}$ rational numbers, defined as numbers of the form p/q where $p, q \in \mathbb{Z}$ and $q \neq 0$

$\mathbb{R}$ real numbers, erm ... you'll have to wait until you're older for a definition!

$\mathbb{C}$ complex numbers, numbers of the form $a + ib$ where $a, b \in \mathbb{R}$ and $i = \sqrt{-1}$

$|x|$ modulus of x

$\overline{x}$ complex conjugate of x

$[a, b]$ see page 8

(a, b) see page 8

f^{-1} inverse function of f or used for preimage set

s.t.   such that

$\pm$ plus/minus

$:=$ defined to be

How to prove that …

It is impossible to give an algorithm that will prove any statement. However, in some cases there are strategies that we can pursue first. In this appendix we give a summary of how to prove various types of statements. Examples are eschewed in favour of brevity.

How to prove that one statement implies another

To prove that A implies B try a number of methods.

- A direct sequence of implications signs.
- Prove the contrapositive statement: prove that the negation of B implies the negation of A.
- Contradiction: Assume the statement is false and prove that this leads to an absurd statement, such as $0 = 1$.

There are many other methods.

How to prove that two statements are equivalent

To prove that A is equivalent to B split the problem in two: prove separately that

A implies B
and
B implies A.

How to prove that two objects are the same

Use some additive or multiplicative structure to show that their difference is trivial. See the following tips on showing two numbers, two sets, are equal.

How to prove that two numbers are equal

To show that $a = b$:

- Prove that $a \leq b$ and $a \geq b$ separately, or
- Prove that $a - b = 0$.

How to prove that two functions are equal

To show that $f(x) = g(x)$ for all x:

- Prove that $f(x) \leq g(x)$ for all x, and $f(x) \geq g(x)$ for all x, separately, or
- Prove that $f(x) - g(x) = 0$ for all x.

How to prove that one set is contained in another

To prove $A \subseteq B$ for two sets A and B:

- Work at the level of set descriptions and use a sequence of equals, or
- Work at the level of elements: Prove '$x \in A \Longrightarrow x \in B$'. Start with 'Let $x \in A$' and proceed from there.

How to prove that two sets are equal

To prove $A = B$ for two sets A and B split the problem into two: prove $A \subseteq B$ and $B \subseteq A$ separately.

How to prove that something exists

To show an object exists, say a certain number, function or set with a desired property:

- Construct it directly, or
- Use proof by contradiction: assume the object does not exist.

How to prove that something doesn't exist

Like the above, assume it does exist, and use proof by contradiction.

How to prove that something holds for all $x \in X$

It is difficult to give general advice in this situation. Many proofs begin with 'Let $x \in X \ldots$' That is, choose an arbitrary element of x that is then fixed throughout the proof.

Try dividing the problem into cases (this is also known as the method of exhaustion). But be aware that you need to do *all* cases; just taking one specific case is not enough. Remember – one example does not prove the general.

Proving with quantifiers

With reference to the three previous bits of advice:

(i) For all x, $P(x)$: You have to show that if someone gives you any x, then $P(x)$ is true.
(ii) There exists x, $P(x)$: You have to find an x, so that $P(x)$ is true.

How to prove that something is unique

Assume that there is another, distinct, one and then show that the two objects are the same, thus producing a contradiction. To do this see the tips above such as showing that their difference is trivial.

How to prove that a set is infinite

To show that a set is infinite:

- Set up a bijection with an infinite set, e.g. $\mathbb{N}$ or $\mathbb{Q}$, or
- Assume the set is finite and apply a contradiction argument. (For example, a finite set of numbers may have a largest element. Show that this gives a contradiction.)

How to prove something indexed by the natural numbers

To prove an assertion with a natural number index use induction.

How to prove something indexed by the integers

To prove an assertion $A(x)$ indexed by the integers, prove it for the natural numbers, for example by induction. Then, for negative x, we have $-x$ is positive, so use $A(-x)$ is true to prove $A(x)$ is true.

How to prove a map is injective

To show that $f : X \to Y$ is injective: assume that $f(x_1) = f(x_2)$ and proceed to show that $x_1 = x_2$.

How to prove a map is surjective

To show that $f : X \to Y$ is surjective: for all $y \in Y$ show that there exists $x \in X$ with $f(x) = y$.

How to prove a map is bijective

Prove it is injective and surjective.

Index